The Innovation

Journey and Outcomes for the AIOps Revolution

ScienceLogic

FREILING
AGENCY

Published by Freiling Agency, LLC.

P.O. Box 1264
Warrenton, VA 20188

www.FreilingAgency.com

PB ISBN: 979-8-9893013-4-8
HB ISBN: 979-8-9893013-3-1
eBook ISBN: 979-8-9893013-5-5

Printed in the United States of America

Contents

Foreword

AS I REFLECT on the remarkable journey that has brought us to this moment, I extend my sincere appreciation to the hundreds who have worn the ScienceLogic badge. To our esteemed, friends, family, and valued customers, your unwavering support has been the cornerstone of our success.

To our loyal customers, your trust and support have been instrumental in our growth, and we are grateful for the privilege of serving you. To our friends and family, your encouragement and understanding have been the foundation upon which we built this business.

Thank you for being an integral part of our journey.

—Dave Link, Founder and CEO, ScienceLogic

1

Introduction

AFTER SPEAKING WITH hundreds of entrepreneurs over the past 20+ years, I have learned there are many catalysts in life that create the impetus for inspiration and innovation. Plato famously wrote, "Our need will be the real creator," which was molded over time into the English proverb, "Necessity is the mother of invention." This is commonly used as the initial idea of germination, but in the case of ScienceLogic it was a unique set of conditions and a moment in time that originated our lightbulb moment.

One of my favorite inventors and entrepreneurs of all time was Thomas Edison, who ultimately brought us many products that changed our way of life, such as the lightbulb, the electric utility system, recorded sound, motion pictures, and the alkaline family of storage batteries. Edison once said, "I have not failed 10,000 times—I've successfully found 10,000 ways that will not work."

As I look back on my career and life's journey, I have always had a fascination for building, creating, experimenting, and then ultimately attempting to fix and improve. This life's motion started early with my father as we were working on carpentry projects in his workshop. This led to building a huge model train layout with sophisticated electrical wiring for lights within multiple train tunnels in mountains we created. Often lights flickered or tracks did not transmit electrical current properly, and required painstaking hours of trial and error to get every light to work at once. As time passed, we moved on to other projects such as the Healthkit electronic processing boards that we built.

Multiple experiments ultimately led to a mini ham radio station that connected to other operators all over the world. These small observations on improving, fixing, enhancing, and creating proactive maintenance routines advanced into fixing broken lawnmowers, dishwashers, household electronics, and thinking through a way to prevent mechanical or structural/ electrical failures before they ultimately happened.

I became very observant looking at everything—old cars, broken-down tractors, and houses and buildings in disrepair. Then I obsessed about where to spend time on each potential project to prevent mechanical or structural failures. As it turns out, my other cofounders independently shared a similar fascination, with Rich growing up on a farm in the United Kingdom where he ran heavy machinery while he proactively cared for the livestock. And Chris, who had an uncanny ability to create, fix, and restore just about anything he put his hands on.

Ultimately, Chris, Richard, and I were going down our own independent career path, originally working in the service provider industry for different Tier 1 carriers with global networks, datacenter infrastructure, and OSS/BSS systems that supported millions of users, customers, and businesses across the globe. Much of our early careers we were focused on delivering IT telecommunication services that spanned the globe.

After many years of independently working within these networks—originally X.25/X.75 networks for me and then large-scale carrier networks, we continued to observe a consistent trend that often created chaos for troubleshooting and delivering service quality that helped the tier 1 & 2 Network Operations team members answer questions at conversation speed in the early stage of system outages.

The core thematic that repeated consistently over the years was the dreaded outage phone bridge that was established upon the initial stages of a service failure and for many was not closed down until the issue was resolved. These bridge conference calls, which could last in the worst of outages 24+ hours, shined a light on some glaring holes in the information gaps that prevented efficient resolution of complex technology services in the early days of company formation.

We all agreed that we felt like at each stage of our career we had undertaken a pathway to deliver great service quality but were often let down by lack of proactive insight into the different technologies that made up the entire service. We will get back to that service layer topic later. Over time we grew increasingly confident that together we had a united view and strong conviction that there was a better way forward.

We had all experienced and run/operated large-scale systems that required many different technologies to come together and act as one unified system to deliver an application outcome to our customers/constituents. Along the way, we had built up highly customized, internally developed, and custom-coded integrations across commercial products to stitch together a unified NOC view.

When we started the company, the network team had their tools, the server team had their tools—the operating systems team had unique tools, the storage database/application tools, virtualization, and ultimately the security and cloud teams had their unique tools. This created a series of stovepipe telemetry and alerting that were not unified—exacerbating these long call bridges (sometimes 48+ hours long) that were initiated during outages. The calls set up like a "who's not it" round-robin conference discussion.

This was super inefficient, verbally going across each team on the call bridge and working to find a root cause of a problem which each technology genera would say, "My systems look fine." In the early days it was the network team that had to prove they were not the problem. However, the cheese has moved and more often this is no longer the case that the network team is first to blame across complex hybrid cloud systems.

Our approach would be totally different in the early days of company formation. Our business conviction and tenacity led us to find clever ways to make each dollar go several miles. We realized that customers were frustrated and tired of 12-16 point tools (often pushed by manufacturers of each technology silo), and NOC users and IT executives craved a unified view. At that time of our company formation, the Big 4 in ITOM (IBM, CA, BMC, and HP) had created some unified tools that were incredibly complex and difficult to deploy as well as run/operate. So, when we started scratching out the optimal path forward in the early days, we wanted to create a new novel approach and apply for intellectual property protection.

We did that within our first year and realized that the fastest time to value was to deliver our ITOM platform as delivered extremely rapid first time to value as we removed the typical installation headaches of installing/configuring the operating system, database, and then ITOM application software on top across a series of servers. We also knew that we needed to deliver a carrier class option for the telco, service provider, enterprise, and federal markets, so we interviewed dozens of potential customers and ultimately assembled a comprehensive list of 240 uniquely differentiated capabilities beyond the table steak features/functions that are expected in a base product.

The initial vision was now complete—create a platform.

Make UI of the platform easy to install and intuitive to operate (eliminate installing OS, DB, application software separately)—just give it an IP address and ethernet connection and you are ready to kick off a discovery and learn some fascinatingly insightful information about your global systems/network environment across what became thousands of heterogeneous technologies. Last, make it easy and smart to upgrade all the components of the platform at once—which was a huge problem in the early days of ITOM platforms.

It all starts with a clear vision of the product/platform that inspires us to engineer the right architectural solutions to achieve our vision. In the early days, this vision to automatically discover the network, servers, operating systems, firewalls, storage arrays, and dozens of supporting technologies was fraught with challenges. Richard spent many sleepless nights perfecting our discovery engine, which ultimately embedded dozens of methods to determine what we had found on the network, how we wanted to classify it, and how to align it to a dependency map and service view. So many times, when we were really stuck without a solution to a seemingly impassible technical blocker, Chris and Richard would identify a compelling and ingenious new method to intelligently leapfrog our blockers.

Over time, intensity of rigor and detailed analysis translated the evolving product platform vision into the detailed architectural designs to support the use case scenarios and unique market opportunity presented in our immediate future, as global service providers started breaking all scale assumptions we had previously estimated.

What does this look like, exactly? It's taken different forms over the years. We've always pioneered IT operations at the very cutting edge, but the world of technology continues to advance at breakneck speed. That is why we have to materially enhance the platform and capabilities every five years.

As an example, after starting the company during the mass datacenter migration from dedicated servers farms to virtualized service farms, running at times a heterogeneous set of hypervisors. The next pivot was now clear—when the compute architecture paradigms change, but management tools and surrounding technologies generally don't make these transitions.

When you think about systems management during the mainframe, these products often did not make the transition from one computing architecture paradigm to the next. These changes are profound and often take the incumbents out of the market opportunity because they don't keep pace to deliver clear business value. In the early days of the company, differentiation was a simple diagram that when the compute architecture changed from mainframe to client server—the tools did not immediately swap out with a new set of tools that were great at managing MSFT technology stack. This too changed again with the rapid evolution of public cloud and associated microservices, which required a different worldview to proactively support the underlying requirements. Along the way, we knew we would need to innovate faster than the marketplace to add maximum business value for our customers.

In the next chapter of the company's history—we are far from being done improving on our UI/UX design to our commitment to integrating machine learning and artificial intelligence to not just improve, but to completely revolutionize, the world of IT operations. No, this isn't just about correlations and optimizations. It's not about better dashboarding and analytics. This is about predicting the future—about IT "wizardry" and using the almost unbelievable power of artificial intelligence to solve problems that we don't ever have to know existed.

It's a tall order. We get that. And it takes a serious commitment to understanding things all the way down to the basics, and mastering the new language of artificial intelligence. But we'll do it.

Today, ScienceLogic is a global pioneer in the IT operations space. We support IT operations for thousands of the world's biggest brands, implementing and creating solutions on the cutting edge of technological innovation. And nothing is more exciting—to me, and, I think, to our talented team—than being on the cutting edge of the burgeoning AIOps revolution.

Over the years, ScienceLogic has grown and evolved, but our commitment to the nerdy pursuit of excellence remains unchanged. We continue to innovate, explore, and push the boundaries of what's possible in the world of IT. And as we look to the future, and ever further into the world of machine learning and AI, we remain dedicated to building solutions that are not only technically sound but also deeply rooted in the passion that defines the true essence of continuous innovation!

2

Customer Story

"WHERE DO I even begin?"

A breathless question. A call for help, maybe. Five words we've all used before that signal the beginning of a chaotic story or the end of a long journey toward a humbling moment of reckoning.

It was late winter 2019. The VP of infrastructure for one of the world's paramount news media titans—a high-stakes conglomerate brand in a uniquely competitive industry—asked this question in front of a group of IT ops engineers at ScienceLogic who'd promised a revolutionary solution for what had become a looming and very serious problem at his organization.

But first, everyone needed to get on the same page about how we got here.

Founded over 100 years ago, this company (a household brand, but unnamed here to protect industry secrets) had seen unprecedented growth in recent decades. Its revolving door of cutthroat executives had executed countless mergers at a breakneck pace. Product owners had adopted dozens of new reader-facing imprints and digital distribution channels. The marketing department was deploying, and even building in-house, new AI-driven curation and moderation applications to custom tailor media content for highly targeted consumer niches. Because of this commitment to growth and innovation, the brand had steadily emerged as an unrivaled pioneer of digital media in the twenty-first century. Daily readership surged by a staggering 800 percent, and they had shed the cloak of corporate obscurity, transforming into a ubiquitous reader-facing brand with a uniquely influential public voice.

Exciting times!

But for this VP (we'll call him Ravi), things had never been more urgent.

Alongside the brand's meteoric ascent, a labyrinthine complex of systems, applications, and legacy networks hobbled along behind. These systems were the arteries that coursed with the highly targeted media content that was the lifeblood of their success. This was their gold—the deliverable around which so many hundreds of systems, and thousands of employees around the globe, were organized.

That Ravi's team struggled to support these systems was a sore spot for executives, and (too often) a drag on their fast-moving, award-winning content producers.

The head-spinning series of mergers, hailed as the result of ingenious strategy and savvy deal-making, had forged a mosaic of an IT infrastructure that lurked ominously in the shadows. This tangled, hardly tamed web cast a pall on their prospects for sustained growth. Within this intricate tapestry of disparate systems was a set of varying arcane rules, throwing Ravi's engineers into a daily struggle with knowledge transfer across teams and new resources.

Integration woes plagued Ravi's infrastructure, too, which gave rise to frustrating bottlenecks that often caught the attention of anxious executives. Security chasms yawned wide, born of misconfigurations and a woeful lack of data protection savvy. Operational costs continued to surge, driven by decision-makers' unrelenting demand for extra resources and specialized expertise across an ever-growing variety of networks and applications.

Ravi, of course, bore the brunt of this cacophony. Day in and day out, his team grappled with the relentless churn of integration troubleshooting. The luxury of time to focus on long-term stability and optimization remained elusive. His days were filled with a regular cadence of rushed troubleshooting and reactive firefighting.

On top of that, the ominous shadow of downtime loomed ever larger as Ravi's infrastructure struggled to keep pace with corporate objectives. In the media business, the toll of such outages extends far beyond mere financial losses. They gnaw directly at the bedrock of customer trust and loyalty, especially as executives make a push for a business model that depends increasingly on paid subscriptions. Disgruntled clients—the victims of these service interruptions—voiced their discontent in public with a fervor that threatened their brand's reputation.

Concerns of data breaches and compliance nightmares emerged, too, born of the fractured and inconsistent cloud landscape Ravi had inherited. Weak points in his security posture invited attacks, while compliance audits unearthed chasms in data protection practices in an industry that has become the target of new legal and regulatory focus.

Blame too many mergers. Blame lazy network engineers. Blame poor communication between corporate decision-makers and IT operators. Blame whoever you want.

But Ravi wasn't here to blame. He was here to solve the problems. If they had been left unchecked, these problems would dismantle the very heart of their mission: the meticulous delivery of tailored news media content to a sprawling readership across an ever-evolving spectrum of platforms.

"Where do I even begin?" A good question, indeed.

Among IT operators, Ravi's problem is all too common. And more so now than ever before.

Over the last decade, the world of IT operations has seen significant changes. It's become more intricate and multifaceted. This transformation has been driven by rapid technological advancements, shifting business requirements, and the relentless march of digital innovation.

Consider cloud computing. Over the last decade, cloud technology has become an indispensable part of IT operations for businesses of all sizes. Amazon Web Services (AWS), one of the leading cloud providers, has grown exponentially since its inception, offering a vast array of services to help organizations scale, innovate, and optimize their operations. This shift to the cloud has not only revolutionized how IT infrastructure is managed, but it has also opened up new avenues for businesses to embrace flexibility and cost-effectiveness in their operations.

Data analytics is another example. The explosion of big data and the advent of powerful analytical tools such as Apache Hadoop and Apache Spark have completely reshaped IT operations. For example, Netflix relies heavily on data analytics to curate content and personalize recommendations for its users. By harnessing vast amounts of data and applying advanced analytics, the company has transformed how it delivers content, creating a more tailored and engaging user experience. This illustrates how IT operations have evolved to encompass infrastructure management and data-driven decision-making, impacting the core of business strategies.

Furthermore, the rise of e-commerce giants such as Amazon and Alibaba has dramatically changed business requirements. These companies have not only disrupted traditional retail but have also reshaped the logistics and supply chain landscape. IT operations play a pivotal role in ensuring seamless order processing, inventory management, and delivery tracking. The need for robust and scalable IT systems to support these operations has never been more critical.

The emergence of blockchain technology, as seen with Bitcoin and other cryptocurrencies, has presented both challenges and opportunities for IT operations. The decentralization and security features of blockchain have raised questions about how to adapt existing IT infrastructures to incorporate this innovative technology. Simultaneously, financial institutions are exploring blockchain for its potential to enhance security and streamline operations.

So now imagine you're leading an IT operations department. How can you summarize all of this for your team? What conceptual apparatus can help them to incorporate the major themes and key trends behind their ever-evolving daily tasks into better ways of thinking and of getting things done and into better plans for scaling bigger and moving faster than ever before?

Here are ten meta-trends to put some context around these changes. This list might not be exhaustive, but it's darn close:

1. **Rapid Technological Advancements:** The pace of technological change has been nothing short of breathtaking. Emerging technologies such as cloud computing, artificial intelligence, the Internet of Things (IoT), and edge computing have introduced new layers of complexity. IT operations teams must now grapple with integrating these technologies into their existing infrastructures while ensuring they remain agile and responsive.

2. **Hybrid and Multi-Cloud Environments:** The shift toward hybrid and multi-cloud environments has introduced a new level of complexity. Managing applications and data across on-premises data centers, private clouds, and multiple public cloud providers requires intricate orchestration and monitoring to ensure seamless performance, security, and compliance.

3. **Security Challenges:** Cybersecurity threats have grown in sophistication and frequency. IT operations teams must manage complex security measures and proactively protect against data breaches, ransomware attacks, and other malicious activities. This necessitates continuous monitoring, threat detection, and rapid response capabilities.

4. **Scalability and Elasticity:** Scalability and elasticity are no longer optional but essential components of IT operations. As businesses grow and digital demands fluctuate, IT teams must ensure that their infrastructure can seamlessly scale up or down to meet these dynamic needs, all while optimizing costs.

5. **DevOps and Continuous Integration/Continuous Deployment (CI/CD):** The integration of DevOps practices and CI/CD pipelines has redefined IT operations. The need for rapid software development and deployment has blurred the lines between development and operations, requiring a cultural shift and the implementation of automation tools and processes.

6. **Data Explosion:** The exponential growth of structured and unstructured data has created storage and management challenges. IT operations teams must handle massive datasets, implement robust data governance, and leverage advanced analytics to extract actionable insights.

7. **Compliance and Regulation:** Regulatory requirements have become increasingly stringent across various industries. IT operations must ensure that systems and data meet compliance standards such as GDPR, HIPAA, or industry-specific regulations, adding layers of complexity to their responsibilities.

8. **Remote Work and Digital Transformation:** The COVID-19 pandemic accelerated digital transformation efforts and ushered in the era of remote work. IT operations teams had to rapidly adapt to support remote infrastructure, address new security concerns, and ensure uninterrupted business operations.

9. **Monitoring and Analytics:** With the proliferation of technology touch-points, IT operations teams rely heavily on advanced monitoring and analytics tools. These tools help identify performance bottlenecks,

predict issues before they occur, and ensure optimal resource utilization.

10. **Skill Set Evolution:** To navigate this evolving landscape, IT professionals have had to up-skill and acquire expertise in such areas as cloud management, cybersecurity, automation, and data analytics. Continuous learning is now a cornerstone of IT operations.

Finally, the COVID-19 pandemic accelerated digital transformation efforts in some drastic and unexpected ways, forcing IT operations to adapt to new remote-first work infrastructures and address a new breed of security concerns.

What does all this mean for teams like Ravi's and for the tens of thousands of diverse IT operations teams around the world?

For one, it means that continuous change must be a fundamental, expected, and ingrained part of daily operations—truly a *way of life*—for IT teams that are committed to fostering innovation and propelling their business toward success. In today's world, innovation is not a one-time, granular, or even sequential event. It's not limited to specific boundaries or initiatives. Rather, it is an ongoing and perpetual process of two steps forward and one step back, fueled by the collaborative efforts of both internal IT professionals and external stakeholders. These processes are sometimes synchronous and in pursuit of defined goals, and other times they are asynchronous and happened upon by chance.

However, it happens, technological innovation is the driving force behind sustainable business growth and development, generally.

This means that corporate decision-makers have no choice but to embrace technology itself as a strategic thinking partner in engineering new and better ways of doing things.

Let's get back to Ravi. After an intense, multidisciplinary effort to scope out the technical specifications of Ravi's unique challenges, ScienceLogic engineers presented him with a plan to implement a new AI-driven platform. The plan was good, and they joined Ravi in pitching this upward—to his executives and board members—to secure the financial and psychological buy-in he needed to make this project a success.

In a matter of months, ScienceLogic had deployed SL1 on premises at Ravi's organization, and the results speak for themselves:

1. **Synchronized Global Operations:** Through the SL1 SaaS platform, a symphony emerged on Ravi's team, resonating across his brand's diverse global coordinates. The transformation of their Network Operations Center (NOC) from a physical realm to a virtualized domain was no minor feat. It required an intricate and piecemeal consolidation process, weaving together teams, processes, and tools into one seamless fabric of IT operational efficiency. As the orchestrator, ScienceLogic's SL1 platform conducted real-time data harmonization, enabling the discernment of impact, the prioritization of actions, and the delineation of responses as an art form. Incidents ceased to be moments of turmoil, transitioning into opportunities for proactive decision-making—a herald of an era characterized by continuous

enhancement through the metamorphosis of digital landscapes.

2. **Illuminated Service Ecosystems:** Analogous to the intricate vines of a jungle, Ravi's organization's business growth introduced a complexity that demanded adept and agile IT management. This led to a shift from conventional *device-centric infrastructure monitoring* to a harmonious symphony of *business service* monitoring. SL1's business services and dashboards unveiled a wealth of proactive insights for Ravi, shedding light on how IT intricately permeated his company's critical business pathways. The outcomes were palpable— superfluous noise was silenced, tasks found alignment, and the cadence of diagnostic processes quickened. This harmony extended beyond SL1, facilitating seamless data flows between other management platforms such as ServiceNow, New Relic, and Sumo Logic. This synchronization, in turn, elevated SLAs, orchestrating responses that exceeded expectations with response times surpassing 90 percent.

3. **Automated Operational Choreography:** SL1's choreographed workflow automation supplanted Ravi's arduous process of identifying, diagnosing, and resolving incidents. This automation composed intricate sequences— synchronizing resources with the Configuration Management Database (CMDB), enhancing tickets with curated triage data, and even orchestrating the rhythm of device restarts.

The culmination of these efforts bore fruit—streamlined and consolidated operations, a palette of real-time insights, a tapestry woven with performance intricacies, and an automated overture leading to efficient IT operations.

In Ravi's own words, "We've been able to simplify and consolidate our operations, achieving real-time visibility and better performance insights, and an automated approach to IT operations."

The metrics tell the story:

1. Elimination of 12 tools
2. A 39 percent reduction in tickets handled by IT staff
3. A 90 percent acceleration in incident response
4. A 16 percent reduction in offshore staff

The journey undertaken by this media colossus alongside ScienceLogic's SL1 was no ordinary (or easy) transformation. It was tough to pull off, but the results resembled a symphony in which innovation, efficiency, and technology harmoniously resonated, orchestrating a reimagined future for Ravi's and his executives' collective imagination.

Now, you may wonder, what were the exact solutions used to drive these changes? We sort of skipped that, didn't we?

Well, here's that story. We'll start with some context. Buckle up!

3

The AIOps Revolution

"THERE'S A LOT of pressure on IT."

That's Dave Link, founder and CEO of ScienceLogic, speaking at the Government, Innovation, Strategy and Technology Conference (GIST) in May 2022.

To seasoned IT professionals, this is an obvious truth. They face this pressure every single day. It's part and parcel of the hyper-competitive landscape that is modern IT. It's what drives innovation across the landscape of technology. Frankly, it's hard to imagine the IT world being different.

But coming on the heels of the COVID-19 pandemic, this obvious truth hits a little harder.

The fact is, IT managers today feel more pressure now than *ever before*. Things have changed. And it's IT, in particular, that's borne the brunt of this seismic shift in the way people live and work. The post-pandemic world demands not just a fresh perspective, but also an entirely new playbook for how technology must perform and adapt to meet the ever-evolving needs of our changing world. Can IT fill in the gaps the COVID-19 pandemic left in its wake? Might we never truly "return to normal," or at least to the way things were before? Do the "old" ways of doing IT apply in a post-pandemic world?

"That pressure has caused all of us to really change the way we think about delivering service quality," Link said. "The complexity of architectures and data that live in many different places require a very different way of running and operating IT."

But there a sense in which this sentiment pre-dates COVID-19. In some ways, the need for continuous re-evaluation has been embedded in the fabric of the IT landscape for quite some time—an undercurrent of unease that has steadily gained momentum over the years. Is it really a creation of recent events, or is it in the very nature of what technology means?

Because it's true. This perpetual sense of urgency has roots deep in the very essence of how IT innovation has happened over the decades. The heart of the matter has always been the relentless need to streamline the ever-growing and intricate IT architectures that underpin our digital existence. The demand for seamless, efficient, and high-performing IT solutions has perpetually surged, often outstripping the pace at which systems and even the best IT professionals can adapt.

Consider the intricate web of interconnected devices, networks, and data repositories that make up modern IT ecosystems. These intricate, multifaceted architectures have expanded exponentially in scale and complexity. Yet the developments in user-end performance, the innovations that shape our day-to-day digital experiences, are intricately dependent on the robustness and agility of these IT infrastructures. In essence, the ability for businesses and end-users to harness the full potential of cutting-edge technologies hinges on the capability of IT operations to scale swiftly and evolve intelligently.

The evolution of IT architectures isn't just a technical imperative. It's also a strategic necessity. Ours is an era where technological advancement shapes the very fabric of society. Staying ahead of the curve requires IT to be more than a support function—it must be a dynamic, proactive force for innovation.

But all that said, one is hard-pressed to find an executive who *won't* argue that the COVID-19 pandemic accelerated certain trends and quickly changed our expectations for how much our technology investments—both at home and at work—should be able to deliver. That infrastructure chokepoint has come under a sharper microscope as demands on IT systems continue to pile up.

Seasoned IT professionals remember that this has happened before. Massive adoption of digital business tooling in the 1990s and 2000s was the catalyst for another kind of innovation—cloud computing—that continues in earnest to this day.

The emergence of cloud computing was a watershed moment in the IT landscape. It represented a fundamental shift in how technology resources were provisioned, managed, and utilized. In essence, cloud computing democratized access to computing power and storage capacity on an unprecedented scale. This shift from traditional, on-premises infrastructure to cloud-based solutions unlocked a wealth of possibilities for businesses and individuals alike.

This time around, IT operations teams are dealing with entirely different solutions. Problems about scale and volume persist, and many of those problems remain unsolved, despite our maturing cloud-first environment. But now, a new kind of technology is changing IT operations teams' framework for identifying and solving these problems, and it's unlike anything we've seen before.

"What we've seen," said Link, "is that data volumes, the velocity, the speed at which things change—machine data coming at *terabytes per day*—we need innovative technologies that do the reasoning over these datasets to look for patterns that provide proactive insights on how stay ahead of service disruptions."

If we could go back to 2003, when Link co-founded ScienceLogic in then-sleepy Reston, Virginia, those comments would probably have us chuckling. Technologies that "reason" with us? It sounds like science fiction. In those days, IT operations teams were wrestling with problems of data volume and speed. The evolving paradigms of server technologies and the nascent cloud applications were striving to deliver data at unprecedented speeds and volumes. The prevailing question was not about predictive insights or service disruption prevention; it was how to grapple with this surge of data while keeping the operational machinery in motion.

The idea that IT could extract original insights from these torrents of data and, in a remarkable twist, predict service disruptions and optimize overall efficiency seemed like a fantastical daydream. The notion of technology becoming an active participant in the decision-making process, equipped with the ability to comprehend, deduce, and predict, was more akin to the musings of futurists than a tangible reality.

But fast forward to today, and the landscape is starkly different. Artificial intelligence (AI) and machine learning have moved from the fringes of technological innovation to the epicenter of IT operations. The once-hypothetical notion of machines reasoning with data and generating actionable insights has materialized into a bona fide revolution. Once grappling with data deluges, IT operations teams now find themselves embracing these technologies at a blistering pace. This seismic shift promises to rewrite IT operations' playbook and their contributions to an organization's bottom line.

This revolution isn't just an evolution. It's much more than that. It's a transformative wave that's rippling across industries, reshaping the very fabric of IT and how we all expect technology to work. The potential implications are profound—from service uptime to resource allocation, from predictive analytics to cost optimization. The days of navigating the IT landscape based solely on reactive measures are giving way to a new era of proactive insight and agile adaptation.

In this unfolding chapter of technological history, Link's vision and ScienceLogic stand as living testaments to the power of ideas, their evolution, and their eventual manifestation into real-world applications. The convergence of visionary entrepreneurship, technological progress, and the unrelenting pursuit of efficiency has ushered in a world where AI-driven IT operations was a fantasy to one where AI is securing its place as the prime mover behind a transformation not just of IT operations, but also of the very essence of how organizations function.

It's Called AI Operations or AIOps

First coined by Gartner back in 2016, AIOps represents the fusion of advanced analytics—namely, machine learning (ML) and artificial intelligence (AI)—aimed at automating IT operations. This new approach enables IT teams to keep pace with the rapid demands of modern businesses where speed, efficiency, and agility reign supreme.

At its core, AIOps is the convergence of big data and machine learning, generating predictive outcomes that empower faster root-cause analysis (RCA) and accelerate mean time to repair (MTTR). By providing intelligent and actionable insights, AIOps drives higher levels of automation and collaboration within IT operations, allowing organizations to continuously improve their efficiency and optimize resource utilization.

But AIOps is more than just predictive analytics. It also champions the cause of automation and collaboration within IT operations. By providing actionable insights to IT personnel, AIOps facilitates quicker decision-making and streamlined processes. This, in turn, fosters a higher degree of automation, as routine tasks can be offloaded to AI-driven systems, freeing up human operators to focus on more strategic initiatives.

The synergy between humans and machines is at the heart of AIOps, promising a more efficient and effective IT environment.

As Dave Link said, when technologies "do the reasoning" in an intelligent way, then they can communicate problems and areas for improvement to both human and robotic operators.

By building real-time systems and context-rich data lakes that span the entire application stack, AIOps filters out the noise and empowers modern performance and fault management systems with automation. Those are big promises, but the impact of this technology on IT operations is hard to overstate. Improved time to resolution, greater application reliability, reduced operating costs, expedited digital transformation, and enhanced user experiences are all within its grasp.

However, as with any technological advancement, the road to realization is not without challenges. Integration with existing IT ecosystems, data privacy concerns, and the need for skilled personnel to interpret and act upon AIOps insights are just a few of the many hurdles that organizations—early adopters, especially—will have to navigate.

But when all is said and done, the AIOps revolution will stand as a testament to the innovative spirit of modern IT leaders. And as with other digital revolutions in the past few decades, this one will quickly make yesteryear's IT solutions the stuff of a bygone (and inefficient) era.

But enough with the jargon. If we're going to truly understand the AIOps revolution, we need to start at the beginning...

The Cloud Computing Revolution

Cloud computing has fundamentally transformed the way IT infrastructure is deployed and managed.

Yes, this is a big *duh*.

For industry insiders, this goes without saying. It's the frank truth that the impact of cloud computing over the past two decades cannot be overstated. In many ways, the modern IT team's day-to-day job is, quite literally, cloud migration, monitoring, and management.

But it wasn't always this way. And in fact, it hasn't been long since cloud computing was brand new and, for many network administrators and cybersecurity analysts, a little nerve-racking.

Cloud computing was a true paradigm shift. It completely transformed the way businesses, and even individuals, manage and deploy their digital resources.

How?

Again, this is not news to anyone in the IT space. But for definition's sake, cloud computing refers to the delivery of computing services—including storage, databases, software, analytics, and more—over the internet. AWS, Microsoft Azure, IBM Cloud, Oracle Cloud are just a few of the applications that have made it possible to run and operate even massive-scale IT architectures without owning physical infrastructure.

This is truly democratized technology.

The benefits of cloud computing are obvious. With cloud services, businesses can easily scale their resources up or down as demand fluctuates, allowing for cost optimization and resource efficiency (particularly useful for startups and small businesses). Pay-as-you-go pricing allows companies to pay only for the resources they use, eliminating the need for large upfront capital investments in hardware and software.

Centralized system updates, maintenance, and analytics relieve IT teams from the burden of routine chores and lend more powerful visibility into system performance and areas for improvement.

This list of benefits could go on forever. The case for cloud computing has been definitely made by the market. AWS alone, according to most analysts, will be worth somewhere close to $3 trillion by 2025—less than 20 years after its founding.

But along with these benefits comes new challenges for IT. The cloud massively multiplies how much data the teams can process. But it also gives rise to a new set of problems.

"Discontinued products and services are nothing new," says Pulitzer Prize-winning writer Nick Carr, whose work focuses on the intersection of IT, business, and culture. "But what is new with the coming of the cloud is the discontinuation of services to which people have entrusted a lot of personal or otherwise important data—and in many cases devoted a lot of time to creating and organizing that data. As businesses ratchet up their use of cloud services, they're going to struggle with similar problems, sometimes on a much greater scale."

The bottom line is that with the cloud's increased capacity, IT teams who are serious about improving user experience have only a new set of problems to solve.

Here are some of those problems.

Challenge 1: Distributed and Microservices Architectures

Cloud computing's widespread adoption gave rise to distributed and microservices architectures, ushering in a new era of modern IT.

The fundamental principles of these architectural patterns involve breaking down monolithic applications into smaller, independent services that communicate with each other over a network. This paradigm shift offered myriad benefits that quickly became evident to organizations across various industries.

One of the most prominent advantages of distributed and microservices architectures is their inherent scalability. Unlike traditional monolithic applications that run on a single server and have limitations on scaling vertically (upgrading hardware resources), microservices enable organizations to scale horizontally. Horizontal scaling involves adding or removing instances of individual services based on demand, allowing them to handle increased traffic or workload without disrupting the overall system. This elasticity of the architecture perfectly aligns with the dynamic nature of cloud computing, as organizations can leverage the cloud's vast resources and elasticity to allocate computing power and storage as needed dynamically.

Organizations can achieve improved fault tolerance and enhanced flexibility by embracing distributed and microservices architectures. These architectures are designed to be resilient, meaning that if one service fails, it doesn't bring down the entire system. Each service operates independently, allowing the overall application to continue functioning even in the face of failures. This high level of fault tolerance ensures that organizations can maintain service availability and reliability, critical factors for modern applications where downtime can lead to severe financial and reputational consequences.

Moreover, distributed and microservices architectures encourage a more agile and iterative development approach. Development teams can work on smaller, manageable codebases, enabling faster deployments and updates. This modularity facilitates continuous integration and continuous delivery (CI/CD) pipelines, making it easier to implement changes, fix bugs, and roll out new features quickly and efficiently.

Despite the undeniable benefits, the adoption of distributed and microservices architectures also introduces new challenges in terms of monitoring and management. Traditionally, monitoring tools and approaches were tailored for monolithic applications residing within the confines of a single server. They thrived in this environment, offering insights into performance metrics, resource utilization, and error rates. However, the transition to distributed systems introduces an entirely new dynamic, where the traditional tools often fail to keep up. The complexity of distributed architectures, with their myriad services communicating across networks, can render these tools inadequate in the face of unprecedented scale.

Monitoring a distributed system calls for a paradigm shift in monitoring practices. It requires a comprehensive view into the performance and health of each discrete service. These services, which operate independently yet interdependently, form the building blocks of the overarching system. It's not just about tracking the performance of each individual service; it's about grasping the intricate web of interactions and dependencies that weave the system's fabric.

To truly comprehend the system's behavior, metrics such as response times, throughput, error rates, and resource utilization must be tracked and aggregated for each service. This mosaic of data offers insights into the ecosystem's holistic health, enabling IT teams to assess not only the performance of individual components but also their collective impact. It's akin to understanding the role of each instrument in an orchestra and how it contributes to the symphony's grandeur.

However, the distributed nature of these architectures introduces another layer of complexity. Issues can transcend the boundaries of individual services, manifesting as anomalies across multiple facets of the system. This necessitates robust tracing capabilities, akin to following a trail of breadcrumbs across the digital expanse. Effective troubleshooting and diagnosis hinge on the ability to trace the path of a request as it traverses the intricate network of services. This tracing not only reveals the origin of an issue but also sheds light on the domino effect it triggers.

In this landscape, monitoring and management morph from static checkpoints into dynamic journeys. They involve not just metrics and dashboards but also adaptive tools that can navigate the labyrinthine network of services. The shift toward distributed and microservices architectures underscores the importance of embracing monitoring tools that can not only contend with complexity but also thrive within it. These tools stand as sentinels, guarding against disruptions, optimizing performance, and providing the insights needed to steer the ship of software architecture through uncharted waters.

Tracking and aggregating metrics such as response times, throughput, error rates, and resource utilization for each service is critical to understanding the system as a whole. And the distributed nature of the architecture means that issues can manifest across multiple services, requiring robust tracing capabilities to diagnose and troubleshoot problems effectively.

Challenge 2: Containerization and Orchestration

The emergence of containerization and orchestration tools has significantly transformed the landscape of modern IT systems, revolutionizing the way applications are developed, deployed, and managed.

Containers, lightweight and isolated by nature, have brought unprecedented portability and consistency to the application deployment process. These "portable" units encapsulate everything needed to run an application, including code, runtime, libraries, and dependencies, making it possible to run the same application reliably across various environments, from development to production.

The benefits of containers are numerous and profound. They allow developers to create applications in a self-contained manner, eliminating the dreaded "it works on my machine" scenario. Consistency in the development and production environments leads to fewer deployment issues, making the application lifecycle more predictable and manageable. Containers also facilitate faster and more efficient scaling, as new instances can be spun up rapidly to meet changing demands. This scalability not only enhances application performance during peak times but also optimizes resource utilization during periods of lower demand.

However, to fully realize the potential of containerization, orchestration tools play a vital role. They provide a robust framework for managing containerized applications at scale, automating various essential tasks and ensuring the seamless operation of complex microservices architectures. With container orchestration in place, organizations can achieve greater agility and efficiency in managing their applications. Auto-scaling features dynamically adjust the number of running containers based on traffic and resource usage, ensuring optimal performance while minimizing operational costs. Load balancing mechanisms distribute incoming requests evenly across container instances, preventing any single component from being overwhelmed. Service discovery tools simplify the process of locating and connecting to different microservices within the container ecosystem, promoting modularity and seamless communication between services.

Despite the undeniable benefits, containerization and orchestration introduce new challenges in terms of monitoring and observability. The dynamic and transient nature of containers makes it difficult to monitor applications consistently. Containers are frequently deployed, scaled up or down, and terminated based on demand. This constant state of flux creates a highly dynamic ecosystem, where traditional monitoring approaches may fall short.

In the traditional monolithic architecture, monitoring applications involved tracking a few well-defined hosts. However, in a containerized environment, the number of containers running at any given moment can fluctuate rapidly, making it challenging to keep track of them manually. IT teams are tasked with monitoring hundreds or thousands of containers, each with its unique set of metrics and logs. As a result, monitoring tools must adapt to this new paradigm, offering scalable solutions that can keep pace with the dynamic nature of containerized applications.

Furthermore, distributed systems and microservices architectures, which are often implemented with containerization, add another layer of complexity to monitoring. Requests may span multiple services, and issues can occur at any point along the service chain, making it crucial to trace and diagnose problems efficiently. Traditional monitoring tools may struggle to provide the holistic view required to identify and resolve these intricate inter-service issues.

Challenge 3: Hybrid Clouds

The allure of scalability, flexibility, and reduced infrastructure costs motivated organizations to dive headfirst into the cloud, eager to harness its transformative potential. In their pursuit of staying competitive and meeting customer demands, many companies took an agile approach, hastily lifting and shifting existing applications into the cloud, while simultaneously developing new cloud-based apps.

The prevailing mindset was to move quickly and capitalize on the latest trends and cost efficiencies without fully comprehending the intricacies and implications of integrating diverse cloud environments. Multiple cloud providers were engaged in a bid to keep up with the ever-evolving landscape, with each provider offering unique services and benefits that seemed irresistible at the time.

IT administrators were caught in the crossfire of executive pressure and high customer expectations. They were expected to deliver the fastest and shiniest new features, often prioritizing speed over thorough planning. Under such circumstances, it's challenging to blame IT teams for their approach to cloud adoption.

However, the aftermath of this hasty cloud rush revealed significant drawbacks. The lack of consistency across cloud platforms resulted in a patchwork infrastructure that IT teams were tasked to manage. Each cloud provider operated with its own set of rules, making it challenging to seamlessly integrate and manage applications across these disparate environments. IT administrators grappled with the complexities of maintaining connections, orchestrating data flow, and ensuring seamless communication between various services.

The consequences of these rushed and fragmented implementations soon became apparent. Integration issues plagued the infrastructure, leading to inefficiencies and bottlenecks. Security vulnerabilities emerged as a result of misconfigurations and inadequate data protection practices. Operational costs skyrocketed due to the need for additional resources and specialized expertise to manage the multi-cloud environment effectively.

The strain on IT teams was palpable, with professionals finding themselves in a perpetual state of integrations troubleshooting. The lack of time to focus on long-term stability and optimization led to a cycle of reactive firefighting rather than proactive innovation. As a result, the promised benefits of cloud computing, such as enhanced cost-efficiency and streamlined operations, appeared elusive, overshadowed by the mounting challenges and complexities.

Downtime incidents became more frequent as the infrastructure struggled to keep up with demand. The consequences of these outages were not limited to financial losses; they also impacted customer trust and loyalty. Disgruntled customers who experienced service disruptions were quick to voice their dissatisfaction, tarnishing the company's reputation and potentially leading to customer attrition.

Even more concerning were the emerging concerns around data breaches and compliance issues. The fragmented and inconsistent cloud infrastructure created weak points in the organization's security posture, making it susceptible to malicious attacks. Compliance audits revealed gaps in data protection practices, leaving the company exposed to legal and regulatory repercussions.

As organizations began to evaluate their multi-cloud strategy, they faced the sobering realization that the initial cost efficiencies were overshadowed by the hidden expenses associated with managing multiple clouds. The additional complexity demanded more manpower, specialized training, and integration efforts, eroding the anticipated savings.

Challenge 4: Increased Demand for Reliability and Uptime

In response to the increasing demand for high reliability and uptime in cloud environments, monitoring solutions must account for the unique challenges posed by distributed, dynamic, and elastic workloads.

But as organizations continue to rely on the cloud to power critical workloads, the demand for and expectations of these monitoring solutions will only increase. The ability to detect, diagnose, and respond to issues promptly is a crucial aspect of ensuring high reliability, uptime, and user satisfaction in today's dynamic and fast-paced cloud computing landscape.

Real-time monitoring in cloud environments is crucial because workloads are no longer confined to a single server or data center. Instead, they are spread across various regions, availability zones, and even cloud providers. This distributed nature of cloud workloads presents a complex and interconnected network of services that must function seamlessly together to ensure optimal performance.

This gets complex.

Modern cloud monitoring tools must employ sophisticated techniques to collect and analyze data in real time from diverse sources, including application logs, performance metrics, and network data. As cloud environments become ever-diversified patchworks of solutions across vendors and environments, it can be challenging—even impossible—to gather and correlate data effectively. For example, an organization might have microservices deployed on one cloud provider and serverless functions on another, all integrated with on-premises legacy systems. Each of these components generates its own set of data and metrics, making it crucial for monitoring solutions to have the capability to aggregate and correlate information from these disparate sources.

Furthermore, with the increasing complexity of cloud-based applications, the number of monitoring data points and metrics can quickly become overwhelming. Modern cloud monitoring solutions need to handle massive amounts of data in real time while maintaining low latency to deliver timely insights. This necessitates the use of high-performance data processing engines and distributed architectures that can efficiently handle the data streams and provide actionable intelligence to IT teams.

Cloud monitoring also requires a multidimensional approach, as it's not just about tracking the performance of individual components but also about understanding the relationships and dependencies between them. The ability to visualize these relationships and gain a holistic view of the entire ecosystem is critical for identifying potential bottlenecks, analyzing service interdependencies, and diagnosing issues that might span multiple services or regions.

Moreover, monitoring solutions must adapt to the dynamic nature of cloud environments. Resources are created, scaled, and terminated on demand in response to fluctuating workloads, making the infrastructure highly elastic. Traditional monitoring tools that rely on static configurations struggle to keep up with these dynamic changes. Cloud monitoring solutions must be agile and automatically adapt to the evolving cloud environment to maintain accurate and up-to-date monitoring.

As cloud environments span multiple cloud providers, regions, and data centers, organizations may also face challenges related to data residency and compliance. Certain data might be subject to regulatory requirements that mandate its storage and processing within specific geographical boundaries—GDPR and the Cloud Act.

Modern cloud monitoring solutions need to address these compliance concerns and ensure that data remains in compliance with relevant regulations while still providing a comprehensive view of the entire cloud infrastructure.

With cloud security being a top priority for organizations, monitoring solutions must incorporate robust security features to detect potential threats and vulnerabilities. This includes real-time anomaly detection, log analysis, and integration with security information and event management (SIEM) systems to detect and respond to security incidents promptly.

The AIOps Solution-Revolution!

There is no future of IT operations that does not include AIOps. This is due to the rapid growth in data olumes and pace of change (exemplified by rate of application delivery and event-driven business models) that cannot wait on humans to derive insights. (*Gartner AIOps Market Guide for AIOps Platforms 2021*)

"As IT tries to modernize, move to the cloud, refactor applications, then run-operate in these compound architectures, they're starting to change the way they run and operate," says Link. "AIOps is next-generational tooling that helps these firms 'bubble up' the signal from the noise."

AIOps equips IT operations teams with an array of capabilities designed to tackle the challenges of the digital age. It enables them to assess the business impact of incidents, diagnose root causes, automate incident resolution, and streamline IT workflows. By leveraging AIOps, businesses gain control and visibility over the consumption of cloud resources, and accelerate MTTR through automated incident management and real-time configuration management database (CMDB) updates.

It's important, though, to consider how AIOps is different from other IT data collection solutions that process and make inferences from data. The fact is, most large organizations have been well integrated with comprehensive data collection tools that gather, organize, and even analyze inputs from various sources. These tools—Zabbix, Nagios, Prometheus, Sensu—were, and continue to be, major players in the move toward the cloud.

However, as any experienced analyst or data scientist will tell you, the abundance of data these tools can generate quickly becomes overwhelming. Humans simply cannot manually review and analyze any of this data. Addressing this challenge by adding dashboards, visualizations, and even advanced query tools is just a band-aid solution. Even with these tools, manual intervention and analysis are still required, making the process inefficient. While they can help to continuously detect potential issues or inefficiencies, pinpointing the most critical areas for improvement can be a challenging task that requires analysts to continuously refactor the parameters of their queries and dashboards as their data architectures grow and evolve.

It can often feel like working in reverse—an analyst must navigate through terabytes of incoming data, hoping to identify the highest-priority areas deserving attention and enhancement.

Yes, he or she might identify areas for improvement or new bottlenecks on data speed. But has the analyst worked on the most pressing problem across the entire architecture? Or just the most pressing problem he or she *happens to find?*

This is where AIOps comes in.

AIOps uses artificial intelligence to ingest, sort, correlate, and make inferences from data—much in the way a data scientist might do if equipped with 10,000 times more time and resources. Instead of relying solely on manual efforts to navigate through overwhelming datasets, AIOps streamlines this process, ensuring that analysts can concentrate their expertise on the most pressing and impactful problems. Moreover, AIOps enhances the decision-making process by providing context-aware insights. It goes beyond merely highlighting issues; it also offers deeper understanding and context surrounding each problem, empowering IT teams to take proactive measures and optimize operations.

And with AIOps, the benefits go beyond uncovering and addressing the most apparent issues. Fully deployed AIOps tools can identify hidden patterns and correlations that might lead to problems in the future in ways that no other tool has ever been able to do. By recognizing potential problem areas before they escalate, organizations can significantly reduce downtime, enhance system performance, and ultimately deliver a more reliable and efficient IT environment for both operators and end-users.

Imagine a sprawling, multi-national corporation with its network functioning smoothly, masked behind a façade of stability. Traditional monitoring tools might grant a superficial nod of approval, but beneath this veneer lie hidden anomalies and areas for improvement that those tools just aren't equipped or designed to identify. This is where AIOps comes into play, armed with predictive algorithms. By sifting through layers of information, it can identify these early warnings, allowing organizations to address problems before they escalate.

In our modern landscape where downtime spells financial losses, AIOps offers a shield against these kinds of disruptions. Its capacity to unearth these latent signals empowers organizations to take proactive measures, minimizing downtime, optimizing system performance, and fortifying the overall reliability of IT operations.

However, the story expands beyond technical matters, touching the lives of human operators and end-users. For

IT professionals, AIOps means relief from sudden outages, allowing them to focus on more strategic, forward-looking tasks. As for end-users, the technology ensures glitch-free experiences, nurturing trust and loyalty.

AIOps' influence underscores the power of data-driven technologies, where algorithms construct stability and resilience. But we're painting with a pretty broad brush here. And that's because it's important to understand AIOps in theory—as the convergence of big data with machine learning—before we can see how exactly it can be deployed in specific environments.

That said, here are some real-world use cases that should help illustrate the tangible value of AIOps applications.

SecOps

AIOps has become an invaluable asset in the realm of SecOps, contributing significantly to the security of applications throughout their development lifecycle-from the early stages of development to delivery and deployment.

During application development, AIOps continuously assesses code changes, dependencies, and configurations in real-time. By analyzing the codebase and identifying potential security weaknesses, it helps developers address vulnerabilities proactively, minimizing security risks before the application is even deployed.

As the application moves toward delivery and deployment, AIOps remains vigilant in monitoring the continuous integration and continuous delivery (CI/CD) pipelines. It examines each build and deployment for any signs of security anomalies or deviations from the norm. This real-time analysis enables quick identification and mitigation of security issues, ensuring that only secure code makes its way into production.

During the deployment phase, AIOps takes a proactive stance in identifying potentially exploitable vulnerabilities that may escape human detection. By analyzing vast amounts of data and utilizing machine learning algorithms, it can spot subtle patterns or anomalies indicative of security threats. This includes unusual access patterns, unauthorized changes to configurations, or malicious activities that human operators may overlook.

AIOps also plays a vital role in supporting SecOps engineers in their efforts to combat security threats. It acts as a force multiplier, assisting security teams in analyzing and prioritizing alerts, allowing them to focus their expertise and attention on the most critical security incidents. By filtering out false positives and providing context-rich insights, AIOps helps streamline incident response and reduces the time it takes to detect and address security breaches. In doing so, AIOps not only mitigates the drain on resources caused by chasing phantom alerts but also liberates SecOps engineers to allocate their skills and efforts where they are most urgently needed.

Furthermore, AIOps is not content with merely presenting a pared-down list of alerts; it enriches each with context-rich insights—valuable layers of data that equip SecOps engineers with a comprehensive understanding of each potential incident, its underlying causes, and potential implications. Armed with this intelligence, the SecOps engineers' response is infused with a newfound level of informed decision-making. This, in turn, engenders a response strategy that is not just swift but also surgically precise—a strategy guided by insights that go beyond the surface, unveiling the intricate tapestry of a security incident.

Furthermore, AIOps continuously learns from security incidents and responses, becoming more adept at identifying new and emerging threats. Its ability to adapt and evolve ensures that the security posture of the application remains strong and up-to-date in the face of an ever-changing threat landscape. In fact, the strategic significance of AIOps' ability to adapt and evolve is perhaps most pronounced when viewed through the prism of this threat landscape. The realm of cybersecurity is far from static; it's a dynamic ecosystem where threats mutate, tactics evolve, and vulnerabilities surface with each technological advance. AI SecOps tools construct a security posture that is not confined to a static moment but is instead equipped to navigate an ever-changing environment. It's a sentinel that augments human expertise, offering insights derived from the amalgamation of past encounters. And as AIOps hones its ability to decipher the language of threats, its contributions extend beyond individual incidents, fostering a security ecosystem that is fortified, adaptive, and always a step ahead of potential adversaries.

DevOps

The collaboration between DevOps and AIOps brings about substantial improvements in the efficiency and effectiveness of build-and-deploy pipelines. By harnessing the power of AI-driven automation and analytics, AIOps enhances various aspects of the DevOps process, leading to increased innovation throughput and greater confidence in software delivery.

One of the key benefits of integrating AIOps with DevOps is the ability to address testing and deployment issues automatically. AIOps continuously monitors the CI/CD pipelines, identifying potential bottlenecks, bugs, or errors in real time. When issues arise, AIOps can trigger automatic responses or initiate self-healing processes, reducing the need for manual intervention and speeding up the resolution of problems. This level of automation streamlines the entire pipeline, enabling DevOps teams to achieve faster and more reliable software delivery.

The increased automation provided by AIOps not only expedites the development process but also enhances the resilience of the application. AIOps can proactively detect anomalies and patterns that might indicate potential failures or performance degradation. By identifying these issues early on, AIOps allows DevOps teams to take preemptive actions and ensure that the application remains stable and robust throughout its lifecycle.

Furthermore, AIOps offers valuable insights into the performance of the application and its components. By analyzing data from various sources, such as logs, metrics, and user behavior, AIOps helps DevOps teams gain a comprehensive understanding of how the software is behaving in real-world scenarios. This data-driven approach fosters an agile feedback loop, enabling teams to quickly identify areas for improvement and make data-backed decisions to optimize the application's performance.

AIOps also enhances the capacity for experimentation and innovation within DevOps. With streamlined pipelines and automated testing and deployment processes, teams can confidently try out new features and changes. AIOps continuously monitors the results, offering real-time feedback on the impact of these experiments. This valuable feedback loop empowers DevOps teams to iterate rapidly and innovate with greater agility.

By promoting faster, more reliable software delivery and fostering a culture of continuous improvement, AIOps strengthens the collaboration between development and operations teams. It breaks down silos, encourages cross-functional communication, and aligns the entire organization toward a shared goal of delivering high-quality software at speed.

The appeal of AIOps continues to grow exponentially. Recent studies show that nearly 30 percent of organizations are currently planning significant investments in AIOps, with more than 90 percent expecting to spend as much or more on AI and machine learning in 2023 as they did in the previous year. Driven by demonstrated performance gains, enhanced productivity, and the need to support ever-increasing volumes of remote operations, AIOps is set to propel future IT operations teams, and their entire organizations, into a future of increased reliability and operational efficiency.

"The theme of modernization with AIOps for IT departments has never been more urgent than it is today," Link says.

Two Paths (a quick note)

There are two distinct approaches to building and integrating an AIOps solution into your IT operations system.

The first of these is called the **data-agnostic** approach. The AIOps system itself does not require any specific domain knowledge or understanding of the underlying IT infrastructure. Instead, it relies solely on historical data and patterns to make decisions and predictions. This kind of AI system is designed to identify anomalies, predict potential issues, and automate certain IT operations based on correlations found in the data.

As you might have guessed, a data-agnostic approach is more *independent*. Because data-agnostic systems are designed to be more generic and applicable to diverse IT environments without the need for extensive customization, they are easier to implement across a wide spectrum of use cases. And they can be easier to scale, given certain parameters on the way the underlying databases are built and grow over time.

They typically don't require the same level of configuration and customization as data-aware systems (which we'll get into below).

Key features of the data-agnostic approach include:

1. **Independence from Domain Knowledge:** Data-agnostic AIOps solutions are designed to be more generic and applicable to diverse IT environments without the need for extensive customization or domain expertise.
2. **Scalability:** Since data-agnostic systems do not rely on specific contextual knowledge, they can be more easily scaled to monitor and manage large and complex IT environments.
3. **Reduced Implementation Complexity:** Data-agnostic AIOps solutions can be relatively easier to implement and maintain, as they do not require the same level of configuration and customization as data-aware systems.

But there's a catch.

A data-agnostic approach relies on a team of professionals—usually specialized data scientists—to make sense of the information this approach yields. And it's no surprise that the vast majority of enterprises do not have access to a whole team of data scientists. Further, this kind of work can become highly, and even overly, specialized, such that switching costs become overwhelming once a particular team has a handle on just how an organization's data and information architecture is designed.

Many IT analysts will say that a data-agnostic approach is more *independent.* And that is true, from one perspective. But it's only true when the right team of data scientists is standing by to help give the system context about the specifics of both the data and the surrounding IT environment. Without this human support, data-agnostic systems may not uncover deeper insights or root causes of complex IT issues.

The other, more pragmatic **data-aware** approach is different. It emphasizes the incorporation of domain knowledge and context-specific information into the AIOps system itself. This involves integrating data from various sources, such as logs, metrics, and events, along with IT domain expertise to create a more contextually intelligent system.

This, of course, requires the AIOps system to "understand" the underlying IT infrastructure, the applications, and the specific environment in which the system operates. Data-aware AIOps solutions take into account the relationships between different components, dependencies, and configurations to provide more contextually intelligent insights and recommendations.

Key features of the data-aware approach include:

1. **Contextual Understanding:** Data-aware AIOps solutions can contextualize data by leveraging domain expertise, historical context, and real-time knowledge about the IT environment. This contextual understanding enables more accurate and relevant analysis of events and incidents.

2. **Root Cause Analysis:** By considering the interdependencies and relationships between various components, a data-aware system can conduct more precise root cause analysis, leading to faster incident resolution and reduced downtime.

3. **Customization and Flexibility:** Data-aware AIOps solutions often require customization and configuration to align with the specific IT environment they are intended to monitor. This can make them more flexible and adaptable to unique requirements.

The data-aware approach eliminates the need for extensive data scientist involvement, because these AIOps models have a better grasp of the IT environment and require less human support to provide more accurate and relevant insights. It empowers the AI itself to make more informed decisions and suggestions when it comes to root cause analysis, issue resolution, and resource allocation.

While these systems are generally more complex and may require more intensive implementation and maintenance than data-agnostic systems (especially as the underlying IT environment inevitably changes over time), a data-aware approach empowers organizations to build a common data model enriched with important context to address diverse business challenges more effectively than a data-agnostic model.

And as an organization's IT infrastructure grows, these systems can eliminate the need for additional data science resources.

ScienceLogic embraces a data-aware approach, empowering organizations to harness the power of data without extensive intervention from outside teams. SL1, their cutting-edge AIOps platform, takes data from any infrastructure and processes it in real time, applying context to make it actionable and meaningful. By connecting the dots between operational patterns and automating workflows, SL1 drives organizations toward operational excellence.

Later on, we'll get more specific about SL1. But first, a real-world example...

4

The Essence of Modern
AIOps-enabled Enterprise

THE YEAR WAS 2021. The world found itself ensnared in the clutches of a merciless and far-reaching global health crisis—one that cast a long shadow of uncertainty and trepidation over every corner of the globe. Humanity was confronted with a formidable adversary, one that transcended borders, cultures, and continents, leaving no corner of the planet untouched by its devastating impact.

Lives had been disrupted, economies had faltered, and healthcare systems were strained to their limits. The pandemic not only upended daily routines but also revealed the vulnerabilities of modern society, forcing individuals and communities alike to confront the fragility of their health and well-being.

Obviously, the worldwide health crisis demanded swift and unprecedented responses from governments, healthcare professionals, and citizens. And these institutions' success (and, for that matter, failures) were well documented. We all watched closely to understand how these changes might affect us—how our lives might change, and maybe forever.

But the lesser told story is that of boardrooms all around the world, where corporate decision-makers were faced with the most difficult set of circumstances they had ever encountered.

The world of business, as they knew it, had been irrevocably transformed. Companies of all sizes and industries were confronted with a stark reality: adapt or face the risk of extinction. The pandemic had unleashed a storm of disruption that reverberated through supply chains, customer demand, and the very foundations of traditional business models.

As the global health crisis raged on, corporate decision-makers were compelled to navigate uncharted waters. They were confronted with pressing questions that extended beyond profit margins and quarterly reports. How could they safeguard the health and well-being of their employees while maintaining business continuity? How could they pivot their operations to meet the changing needs and expectations of customers in an era of social distancing and remote work?

In response to the pandemic's challenges, a multitude of sectors underwent transformative changes. Restaurants, for instance, embraced outdoor seating arrangements to accommodate social distancing guidelines while preserving their dine-in experiences. Retailers, recognizing the surge in demand for contactless shopping, expedited the expansion of their online ordering and curbside pickup services, providing customers with safer alternatives. Simultaneously, a significant paradigm shift was observed in the corporate world, as work-from-home arrangements swiftly gained prominence and have continued to do so.

For major corporations boasting vast workforces, the transition to a remote work model placed unprecedented demands on their information technology (IT) infrastructure and resources. This shift presented a unique set of challenges, as companies needed to ensure not only the connectivity and productivity of remote employees but also the security of their digital assets and sensitive information. The sudden surge in remote work exposed vulnerabilities, requiring organizations to bolster their cybersecurity measures, enhance network capacities, and optimize remote collaboration tools.

The airline industry was especially hard-hit.

Derek Whisenhunt, then head of End-User Computing and Enterprise Monitoring Engineering at Southwest Airlines, offered a sobering glimpse into the airline's tumultuous journey during his keynote address at Symposium 2021. He characterized the airline's predicament as "chaotic," a stark contrast to the robust performance it had enjoyed in the year preceding the crisis.

In the pre-pandemic era, Southwest Airlines had been riding high, routinely ferrying an impressive half a million passengers to and from 117 different destinations on any given day. Then, in March of 2020, the industry's fortunes took an abrupt nosedive, and Southwest was suddenly grappling with a passenger count reduced to a mere 22,000.

For Southwest Airlines, renowned for its commitment to "obsessive customer service," the challenge lay in the fact that, although a substantial portion of their nearly half-million-strong customer base was no longer traveling, the customer service lines were abuzz with calls. These calls were primarily related to the need for flight cancellations or rescheduling, requests for refunds, and inquiries regarding policy changes and safety measures. Remarkably, all of this unfolded against the backdrop of Southwest's ongoing efforts to implement a much-needed technological overhaul, crucial for adapting to the "new normal."

Whisenhunt explained the complexity of this endeavor, emphasizing that architectural changes typically involved protracted timelines that spanned one to two years. However, Southwest Airlines found itself confronted with an urgent need for transformation, compressing this timeline into a matter of days rather than years. From an engineering perspective, the challenge was nothing short of Herculean: they needed to transition a company designed for employees to function within the confines of traditional office spaces into a flexible framework that would enable them to work effectively from their homes.

But in the midst of these unprecedented challenges, Whisenhunt and his IT team found themselves confronting an additional hurdle—they were also compelled to work remotely as much as possible. However, rather than merely reacting to the crisis, Whisenhunt saw an opportunity to embark on a significant journey of transformation.

He articulated the team's aspirations succinctly, expressing that their goal extended beyond the preservation of operational continuity; they aimed to modernize their entire IT infrastructure. This ambition set the stage for an ambitious IT transformation initiative within Southwest Airlines.

Whisenhunt, at the helm of Southwest Airlines' IT transformation efforts, articulated a comprehensive and visionary set of objectives that would not only steer the airline's IT department towards a brighter future but also ensure that Southwest remained at the forefront of the aviation industry. These objectives formed the strategic foundation for their IT transformation journey, embodying the commitment to efficiency, customer-centricity, and innovation.

1. **Tools Consolidation:** This objective was at the core of the transformation initiative. Southwest recognized that they possessed a multitude of tools and technologies within their IT ecosystem, and the complexity that ensued needed to be addressed. By streamlining their toolkit, they aimed to reduce complexity, cut down on redundancy, and create a more efficient and manageable IT environment. This consolidation would not only save time but also make it easier for their IT teams to collaborate effectively.

2. **Operational Excellence:** Elevating IT operations to a state of excellence (not just "good enough") was paramount. Southwest understood that the quality and efficiency of their IT operations had a direct impact on the overall performance of the airline—and even more so in the pandemic and post-pandemic era. Whether it was ensuring the reliability of critical systems or responding swiftly to emerging challenges, IT operational excellence was going to be the linchpin to their continued success.

3. **Great User and Customer Experience:** Southwest Airlines has a longstanding reputation for exceptional customer service. This objective reinforced their commitment to the passenger experience. By placing a strong emphasis on the end-user and customer experience, they aimed to ensure that technology was an enabler of exceptional service, from booking a flight to the moment a passenger stepped off the plane.

4. **Self-Service Catalog of Services:** Empowering users with self-service capabilities was an integral part of

Southwest's transformation strategy. This allowed passengers, as well as internal stakeholders, to access and manage IT services independently. It not only reduced the burden on IT support but also empowered users to have more control over their interactions with technology, enhancing overall efficiency and satisfaction.

5. **Cost Saving:** While pursuing innovation and improvement, cost-effectiveness remained a critical consideration. Southwest Airlines sought to implement strategies that would optimize IT spending without compromising the high standards of service they were known for. This objective aligned with the broader goal of maintaining financial stability while investing in IT enhancements.

6. **Automation:** Embracing automation was a forward-looking move to drive efficiency and minimize human error. By automating repetitive tasks and processes, Southwest aimed to free up their IT professionals to focus on more strategic, creative, and value-added endeavors. This not only increased productivity but also improved the accuracy and reliability of IT operations.

Southwest Airlines opted for a "greenfield approach" to this transformation—a strategy necessitated by the scale of their IT landscape. With numerous applications and services to support, it became evident that a comprehensive overhaul was the company's most viable path forward. This approach aimed to migrate a significant portion of their infrastructure to the cloud, thereby reducing their on-premises IT footprint to a minimum.

The cloud-centric strategy offered several advantages. It not only enabled Southwest to achieve greater flexibility and scalability but also facilitated the rapid adoption of new products and services as needed. This approach came with the added benefit of assured integration, eliminating many of the compatibility challenges that often accompany major technological transitions. In essence, Whisenhunt wanted to position Southwest Airlines not just to weather the storm but to emerge from it as a leaner, more agile, and technologically advanced organization poised for the demands of the future.

This vision aspiration for a modern and agile IT landscape called for the implementation of a robust IT operations monitoring platform. This platform needed to possess several critical capabilities, including the ability to identify and manage integration components across a hybrid environment, keep pace with the constantly changing nature of IT resources in real time, and create an operational data repository filled with standardized data. This reservoir of information would not only instill confidence in the monitoring team regarding the health, availability, and reliability of resources but also supply the essential data and insights required to facilitate IT process automation.

After careful consideration, Southwest Airlines selected ScienceLogic SL1 as the foundation upon which to construct their new IT environment.

Because ScienceLogic SL1 continuously collects data from various systems and applications, this real-time data can be used to identify and analyze potential issues and anomalies before they could escalate into critical problems. With proactive monitoring, Southwest could ensure uninterrupted services, minimize downtime, and enhance passenger satisfaction.

Southwest also recognized the importance of automation in improving efficiency and reducing the risk of human error. ScienceLogic SL1 platform provided automation capabilities that allowed Southwest to automate routine IT tasks and workflows. For example, ticketing and incident management processes could be automated, ensuring a faster response to IT issues and minimizing the impact on operations.

Whisenhunt emphasized the magnitude of the transformation, highlighting that when they commenced their journey, they were managing an unwieldy array of 33 distinct tools. With the integration of SL1, this complexity was streamlined down to just six tools, a dramatic consolidation that vastly simplified their infrastructure management. Equally significant was the liberation of engineers, analysts, and other teams that were previously preoccupied with maintaining the now-redundant tools. These resources could be reallocated to provide support to customers and the infrastructure and operations teams, amplifying their overall effectiveness.

Southwest Airlines experienced several operational efficiency gains throughout this transformative process. These efficiencies extended beyond streamlined IT operations. Productivity soared with the adoption of collaboration tools, and the elimination of daily commutes translated into increased staff availability and productivity.

In essence, Southwest was reaping the benefits of a more agile and technology-driven work environment.

Looking ahead, the airline giant's roadmap includes the integration of several Software as a Service (SaaS) solutions. Furthermore, they plan to migrate their applications and infrastructure teams into this new environment, ensuring that they are equipped to deploy resources at the departmental level efficiently. The ultimate objective, according to Whisenhunt, is to automate various aspects of their IT operations continually.

Whisenhunt's vision of success lies in the transformation of Southwest's IT operations team from a reactive force to a proactive one. He envisions an era where they can foresee and preempt issues before they arise, responding automatically and predictively. This aspiration encapsulates the essence of modern AIOps-enabled enterprise monitoring—enabling organizations not only to manage their IT assets effectively but also to harness the power of data and automation to anticipate and mitigate challenges proactively.

5

What Is SL1?

SIMPLY PUT, SL1 is ScienceLogic's comprehensive AIOps solution. It does everything (and more) that we discussed in Chapter One—everything an AIOps solution should do.

Let's start in layman's terms.

Imagine ScienceLogic SL1 as a digital guardian for your company's computer systems. Its job is to watch over everything—devices, networks, and software—just like a security guard keeps an eye on a building.

If something unusual happens, such as a device acting strangely or a network getting slow, SL1 notices it right away. It's a bit like how your senses tell you when something doesn't seem right, such as a strange noise or a strange smell. When SL1 notices a problem, it doesn't keep it to itself. It sends a message to the people who can fix the issue. This message is called an alert. It tells them what's going wrong and where.

But SL1 doesn't stop there. It's also clever. It can be programmed to automatically do certain tasks to solve problems. For example, if a computer crashes, SL1 can restart it without waiting for someone to do it manually. This helps in fixing problems quickly, just like a superhero swooping in to save the day.

Cool, right? But those are just words, really. How does SL1 work?

To be precise, SL1 is an advanced IT operations and monitoring platform that empowers businesses with comprehensive visibility and control over their IT infrastructure, applications, and services. It's square in the center of the growing AIOps application landscape and the front-runner in the AIOps innovation race.

Think of SL1 like a seeing stone that gives users (most often, IT operations managers) a portal into the inner-working of *just about everything that matters to them.*

To date, more than 15,000 companies around the world use SL1. We've told some of these stories already, and we'll share more in later chapters. But even as the AIOps revolution is just beginning, SL1 has gained its chops as the world's leading AIOps solution.

SL1's biggest strength is its *unified observability capabilities* (though that's just another way to define SL1 itself). It's literally a centralized hub that collects, consolidates, and correlates data from a wide range of sources—network devices, servers, virtual machines, cloud resources, applications, and more. This unified view provides IT teams and executives with a truly holistic perspective, enabling them to quickly identify and troubleshoot issues that inevitably arise across different environments.

And with SL1, that perspective comes with a real-time metrics dashboard and intelligent analytics. This way, IT teams can identify performance trends, pinpoint data bottlenecks, and optimize resource utilization.

But of course, observability and event management are two peas in a pod. It's worth it to see and identify problems only if you have the power to resolve them. This is why SL1 comes with built-in *event management functionality*. This empowers organizations to not just monitor for incidents but also to efficiently manage and solve them. SL1 analyzes events and alerts from various sources and applies intelligent algorithms to prioritize them according to their criticality. This helps IT teams resolve the highest impact incidents first, meaning faster problem resolution and improved service availability. And ultimately, of course, a better customer experience and bottom-line ROI.

Some of SL1's other features include the following…

Service Mapping and Dependency Visualization

Service mapping and dependency visualization are critical components of SL1's capabilities, empowering IT teams with a comprehensive understanding of their organization's IT ecosystem. These features go beyond traditional monitoring tools, enabling IT administrators to gain valuable insights into the complex relationships and dependencies between various IT assets, applications, and services.

Service mapping, powered by advanced discovery algorithms, automatically identifies and maps the connections between IT components in real time. This dynamic mapping provides a constantly updated representation of the entire IT infrastructure, regardless of its complexity or scale. IT teams can visualize the relationships between servers, virtual machines, applications, databases, network devices, and other components, helping them grasp the interdependencies that define their IT landscape.

The visualization of these relationships and dependencies is instrumental in assessing the impact of incidents, changes, or disruptions, focusing on the bottom-line business impact. When an issue arises, IT teams can quickly pinpoint the affected components and assess how the problem propagates through the system. Understanding the affected services and the applications' underlying infrastructure helps IT administrators prioritize their response efforts, ensuring that critical services are promptly restored.

Root cause analysis, a time-consuming process in many IT environments, becomes significantly streamlined with SL1's comprehensive service mapping and dependency visualization. IT teams can trace the sequence of events leading up to an incident, easily identifying the root cause among interconnected components. The ability to visualize dependencies helps avoid the classic "blame game" scenario, in which different team members point fingers at one another without clear evidence. Instead, a data-driven approach empowers IT teams to collaborate efficiently and focus on resolving the underlying issue promptly.

One of the key differentiators for SL1 is its ability to build accurate and reliable topology maps. SL1 doesn't merely provide individual data points; it constructs a holistic view of the entire IT ecosystem, showcasing how each component relates to others. This comprehensive topology map enables IT teams to spot trends, identify bottlenecks, and make informed decisions about performance optimization and resource allocation.

Moreover, ScienceLogic SL1's visualization capabilities are not limited to current infrastructure snapshots. The platform also maintains historical mapping data, allowing IT teams to analyze changes and trends over time. This historical view can reveal how the IT ecosystem evolved, helping teams identify patterns and proactively address potential issues before they escalate.

Additionally, the ability to visualize dependencies extends to proactive monitoring and capacity planning. IT administrators can anticipate the impact of changes or additions to the IT infrastructure, ensuring that new applications or services are seamlessly integrated without causing disruptions or resource bottlenecks. This proactive approach aids in maintaining high performance and service availability, promoting a smooth and uninterrupted user experience.

Furthermore, SL1's visualization is not limited to a single platform or environment. It supports hybrid and multi-cloud deployments, where components may be distributed across various cloud providers and on-premises data centers. The platform creates a unified view of the entire hybrid IT environment, allowing IT teams to manage and troubleshoot seamlessly across multiple platforms.

By leveraging both historical and real-time data, SL1 offers IT teams a robust set of tools to forecast resource requirements, make data-driven decisions, and drive efficient infrastructure investments.

Historical data is a treasure trove of insights, allowing organizations to analyze past trends and patterns in resource utilization. By examining historical data, IT teams can identify seasonality trends, peak usage periods, and historical performance metrics. Armed with this knowledge, they can proactively plan for future capacity needs and allocate resources optimally to handle fluctuations in demand.

Moreover, SL1's real-time data capabilities provide IT teams with up-to-the-minute visibility into the current state of their infrastructure. Real-time monitoring enables immediate detection of anomalies, bottlenecks, or sudden spikes in resource usage. With this real-time insight, IT administrators can swiftly respond to emerging issues, preventing potential disruptions and downtime.

The combination of historical and real-time data empowers IT teams to conduct thorough capacity planning. Organizations can avoid the pitfalls of over-provisioning, which can lead to unnecessary costs, or under-provisioning, resulting in poor application performance and user dissatisfaction.

Over-provisioning—the act of allocating more resources than necessary—can result in needless expenses. Imagine dedicating excessive computing power, storage, or network capacity to a task that could operate efficiently with less. This not only inflates costs but also ties up resources that could have been better allocated elsewhere. By analyzing historical usage patterns and real-time demands, AIOps equips IT teams with insights to precisely gauge the required resources, sidestepping over-provisioning and its financial implications.

Conversely, under-provisioning can have equally detrimental consequences. Insufficient resources lead to strained systems, sluggish application performance, and disgruntled users. Think of this scenario: a business-critical application grinds to a halt due to insufficient computing power or a lack of storage. The aftermath could involve missed opportunities, hampered productivity, and tarnished user experiences. This is where AIOps shines. By drawing from historical data and real-time inputs, it empowers IT teams to anticipate spikes in demand and allocate resources accordingly, thus averting potential bottlenecks.

Capacity planning isn't just a balancing act; it's a finely tuned orchestration of resources to maintain a delicate equilibrium between costs and performance. AIOps equips IT professionals with a dynamic playbook, allowing them to forecast resource requirements accurately. With this intelligence in hand, they can navigate the terrain of computing power, storage, and network capacity, ensuring a harmonious flow that doesn't buckle under pressure.

The impact of this precision is multifaceted. Not only does it mitigate financial waste and optimize cost-efficiency, but it also bolsters the reliability of IT operations. Systems function optimally, applications hum along smoothly, and users experience the responsiveness they expect. It's akin to finding the sweet spot on a pendulum, where the delicate equilibrium between resource availability and application demand is maintained.

In the grand scheme of IT management, capacity planning is the linchpin that wields the power to shape an organization's technology landscape. The inclusion of historical and real-time data within this strategic endeavor transforms it from a gamble into an informed decision-making process. As businesses navigate the evolving currents of technological demands, AIOps is the compass that guides them toward optimal resource utilization, cost-effectiveness, and operational reliability.

Advanced Analytics and Machine Learning Capabilities

At the heart of SL1's advanced analytics is the ability to correlate data from diverse sources across the IT ecosystem. This includes performance metrics, log files, configuration data, and even user behavior data. By aggregating, correlating, and analyzing this wealth of information in real time, SL1 can paint a comprehensive picture of the IT environment and identify patterns or anomalies that may indicate underlying issues.

The intelligent algorithms at work within SL1 continuously learn from historical data and ongoing patterns, enabling the platform to recognize common incident signatures and detect deviations from normal behavior. These self-learning capabilities make SL1 increasingly effective over time, as it refines its understanding of the IT ecosystem and becomes more adept at distinguishing between benign anomalies and genuine incidents.

The true power of SL1's advanced analytics lies in its ability to swiftly pinpoint the root cause of IT incidents. By analyzing multiple data streams in real time, SL1 can identify the component or process responsible for triggering an incident. This is especially invaluable in complex and dynamic IT environments where the traditional manual investigation can be time-consuming and error-prone.

Accelerating the mean time to repair (MTTR) is a critical advantage of SL1's root cause analysis capabilities. With the root cause quickly identified, IT teams can take decisive action to resolve the issue promptly. The automation and intelligence within SL1 enable it to suggest potential solutions based on past resolutions or best practices, further streamlining the incident response process.

The reduction in downtime achieved through effective automation, preventing issues, and root cause analysis has a cascading impact on service quality and customer satisfaction. Unplanned downtime can lead to substantial revenue losses, damaged brand reputation, and eroded customer trust. By minimizing downtime and swiftly resolving incidents, SL1 helps organizations maintain high service availability and keep their customers happy.

Moreover, SL1's advanced analytics and machine learning capabilities go beyond incident resolution. They also play a vital role in proactive monitoring and predictive maintenance. SL1 can detect early warning signs of potential issues and alert IT teams before they escalate into critical problems. By addressing potential problems proactively, organizations can prevent costly downtime and improve the overall reliability of their IT services.

As data volumes and IT complexity continue to grow, manual analysis and troubleshooting become increasingly impractical. Advanced analytics and machine learning are the future of IT operations, and SL1 is at the forefront of this transformation. With its powerful capabilities, SL1 enables IT teams to harness the full potential of their data, gain actionable insights, and optimize the performance and reliability of their IT infrastructure.

Automation and Orchestration Capabilities

Another benefit of SL1 is its ability to streamline routine tasks through automation.

In traditional IT environments, IT personnel often spend a significant amount of time on repetitive, manual tasks—configuration changes, application deployments, and system monitoring. These tasks are not only time-consuming, but they also expose vital IT functions to significant and consequential human error. SL1's automation capabilities take these mundane tasks off the hands of IT administrators, allowing them to focus on higher-value activities that drive innovation and business growth.

Furthermore, SL1 ensures that these automated processes are executed consistently and accurately. When tasks rely on manual execution, there's always a risk of human error creeping in, which can lead to inconsistencies and even operational hiccups. SL1's automation takes this worry off the table by following well-defined and standardized steps, resulting in predictable and dependable results.

This dependable execution has several tangible benefits. First, it significantly boosts operational efficiency. With SL1 in charge, tasks are carried out with precision, making the most efficient use of resources and time. This streamlined workflow not only saves money but also reduces the need to fix mistakes caused by human errors or handle the fallout from service disruptions.

Moreover, fewer errors translate to a lower likelihood of service interruptions. Manual processes are susceptible to unexpected problems that can disrupt services, causing frustration for customers and financial losses for companies. By replacing these manual processes with SL1's consistent and error-free automation, organizations create a reliable shield against these disruptions, ensuring customer satisfaction and maintaining a strong position in the market.

SL1's automation capabilities extend to remediation actions, too. This allows the platform to automatically respond to incidents based on predefined rules and policies. When SL1 detects an issue, it can trigger encoded remediation actions to resolve the problem promptly. For example, if SL1 detects a server performance issue, it can automatically allocate additional resources or restart the affected service to restore normal operation.

By automating remediation actions, SL1 reduces the mean time to resolution (MTTR) for incidents, leading to improved service availability and faster incident response. This proactive approach to incident resolution helps organizations maintain high service levels, meet SLAs, and deliver an exceptional user experience.

Another area in which SL1's automation shines is in workflow orchestration. IT teams often deal with complex processes involving multiple tasks and stakeholders. SL1's automation can orchestrate these workflows, coordinating and synchronizing the execution of various tasks across different systems and teams. For instance, SL1 can automate the process of provisioning new resources, from requesting approvals to deploying configurations, ensuring a seamless and error-free process.

Automated workflows improve cross-team collaboration and streamline communication, reducing the risk of miscommunications and misunderstandings. With SL1's workflow automation, IT teams can work more efficiently, accelerating project timelines and achieving better outcomes.

Integration and Extensibility

We've alluded to this already, but integration and extensibility are core pillars of SL1, ensuring that organizations can harness the full potential of their existing IT management and monitoring investments.

The ability to integrate with a diverse array of IT management and monitoring tools is essential in today's complex IT environments. Organizations often have existing investments in specialized tools for various tasks, such as network monitoring, application performance management, log analysis, and cloud management. SL1 recognizes the value of these investments and avoids the need for a "rip and replace" approach.

SL1's integration capabilities enable it to pull in data and insights from these existing tools, consolidating the information into a single, unified platform. By integrating with other monitoring solutions, SL1 creates a comprehensive and holistic view of the IT ecosystem, enabling IT teams to have complete visibility into their infrastructure's health and performance.

Furthermore, SL1's integrations extend beyond traditional on-premises solutions to encompass cloud-native tools and services. With the growing adoption of cloud computing, organizations often manage resources across multiple cloud providers and regions. SL1's integrations with leading cloud platforms and cloud-native monitoring tools allow IT teams to gain insights into their cloud infrastructure, ensuring optimal performance and cost efficiency.

A critical aspect of SL1's integration capabilities is its support for APIs (Application Programming Interfaces). SL1 provides a comprehensive set of APIs that enable organizations to develop custom integrations and extend the platform's functionality to suit their specific business requirements. APIs allow IT teams to extract data, initiate actions, and create custom dashboards programmatically.

The flexibility of SL1's APIs enables organizations to build tailored solutions and workflows that align with their unique processes and goals. This extensibility fosters innovation and empowers IT teams to customize the platform to fit their specific needs, enhancing operational efficiency and effectiveness.

Moreover, SL1's customizable dashboards empower organizations to create visualizations and reports that reflect their unique data requirements and key performance indicators (KPIs). Each organization has its own set of metrics and business goals, and SL1's customizable dashboards enable them to focus on the data that matters most to them.

Customizable dashboards also facilitate collaboration across teams and departments. Different stakeholders may require specific views of the data to make informed decisions. With SL1's customization capabilities, organizations can tailor dashboards to meet the needs of various teams, fostering collaboration and alignment.

As the IT landscape continues to evolve, ScienceLogic remains committed to expanding its integration capabilities to keep pace with emerging technologies and trends. SL1's integration and extensibility features ensure that organizations can adapt and scale their IT operations effectively, maximizing the value of their investments and staying ahead in a rapidly changing digital world.

Capacity Planning

By leveraging both historical and real-time data, SL1 offers IT teams a robust set of tools to forecast resource requirements, make data-driven decisions, and drive efficient infrastructure investments.

Historical data is a treasure trove of insights, allowing organizations to analyze past trends and patterns in resource utilization. By examining historical data, IT teams can identify seasonality trends, peak usage periods, and historical performance metrics. Armed with this knowledge, they can proactively plan for future capacity needs and allocate resources optimally to handle fluctuations in demand.

Moreover, SL1's real-time data capabilities provide IT teams with up-to-the-minute visibility into the current state of their infrastructure. Real-time monitoring enables immediate detection of anomalies, bottlenecks, or sudden spikes in resource usage. With this real-time insight, IT administrators can swiftly respond to emerging issues, preventing potential disruptions and downtime.

The combination of historical and real-time data empowers IT teams to conduct thorough capacity planning. Organizations can avoid the pitfalls of over-provisioning, which can lead to unnecessary costs; or under-provisioning, resulting in poor application performance and user dissatisfaction. By accurately forecasting resource requirements, IT teams can ensure that they have the right amount of computing power, storage, and network capacity available at all times, minimizing waste and optimizing cost-efficiency.

A Trip through Nodeville

When you think about SL1, think about nodes.

SL1 is built on nodes. SL1 adopts a distributed architecture that operates using a node-based approach.

Why?

Because nodes naturally provide a scalable, resilient, and high-performance solution. This is as true for IT infrastructure and application monitoring (like SL1) as for any other IT-related task or application.

But how exactly should we think about nodes in SL1? Here's a story to explain:

Imagine a bustling city—Nodeville—where different tasks and responsibilities are distributed among specialized individuals. Each of these people has unique skills and roles. Nodeville works like a charm because its inner workings mirror a node-based architecture, and its organization mirrors the benefits it brings to IT applications.

In Nodeville, scalability reigns supreme. As its population grows and demands on its infrastructure increase, new individuals join Nodeville's workforce seamlessly, allowing the city to expand without overwhelming any single person. The workload is shared among these capable individuals, ensuring that tasks are completed efficiently and promptly. In other words, Nodeville is always ready to handle and integrate more and more residents!

Flexibility and modularity are the pillars upon which Nodeville stands or falls. Each resident of Nodeville has his or her domain of expertise and can adapt to changing circumstances. They're all well trained and can easily swap roles, acquire new skills, or incorporate innovative tools into their work. This adaptability enables the city to stay ahead of the curve, readily embracing advancements and meeting evolving requirements.

Fault tolerance and resilience are Nodeville's backbone. If one person encounters a problem or becomes unavailable, the city doesn't crumble. Backup individuals stand ready to step in effortlessly to carry out the necessary tasks. This ensures that the city's operations continue uninterrupted, even when one part of its workforce becomes strained or overworked. This collective strength and redundancy guarantee that Nodeville remains resilient and responsive, even in challenging situations (say, during a natural disaster).

Efficiency and optimization define Nodeville's work ethos. Each individual focuses on his or her specific responsibilities, honing skills and optimizing performance. All residents develop expertise in their areas, leading to remarkable proficiency and productivity. By leveraging this specialization, Nodeville maximizes its potential, accomplishing tasks swiftly and efficiently.

Just like Nodeville's residents collaborate and work in parallel, the city harnesses the power of parallel processing.

Multiple individuals work concurrently on different parts of complex tasks, ensuring rapid progress. This parallelization empowers Nodeville to handle large-scale projects efficiently, enabling significant speed and improved overall productivity.

Distributed computing is another hallmark of Nodeville. The ability to share resources and collaborate seamlessly allows the city to tackle monumental challenges. Individuals communicate effortlessly, combining their capabilities to accomplish tasks that would be impossible for a single person working alone. This distributed approach unleashes the city's full potential.

Modifiability and extensibility are also deeply ingrained in Nodeville. Individuals continuously adapt to changing circumstances, integrating new knowledge, tools, and techniques into their work. They stay relevant and future-proof by readily embracing innovations and advancements. This flexibility ensures that the city remains adaptable and versatile, ready to tackle emerging challenges head-on.

Just as this thriving city embodies the benefits of node-based architecture, IT applications flourish under its principles. The efficient distribution of tasks, scalability, flexibility, fault tolerance, parallel processing, distributed computing, and adaptability empower organizations to build resilient and high-performing systems. By embracing the spirit of Nodeville, IT architectures can unlock their full potential, achieving remarkable efficiency, agility, and success in a rapidly evolving digital landscape.

Make sense?

Maybe that's a silly example—especially for someone who's worked in-depth with the kind of node-based architecture inherent in SL1. But even still, it's valuable to step back and remember what this is all about—why nodes make sense, how to build upon this architecture, and how to think about ways to enhance the way it works for your organization.

But what do nodes actually look like in SL1 (not in Nodeville)? Let's explore. (Oh, and beware of heavy jargon ahead!)

Collector Nodes

Collector nodes lie at the heart of ScienceLogic SL1's architecture, representing a foundational element that ensures comprehensive and efficient data collection from across the entire IT landscape. By strategically distributing these nodes throughout the IT environment, SL1 can effectively gather data from diverse sources, providing IT teams with a holistic view of their infrastructure's health and performance.

The deployment of multiple collector nodes addresses the challenge of managing vast and distributed IT infrastructures. In modern organizations, IT assets and services are often spread across different geographic locations, data centers, and cloud providers. This distributed nature can create silos of data that hinder the ability to gain a unified and coherent understanding of the overall IT ecosystem.

With SL1's strategically placed collector nodes, data collection becomes a streamlined and coordinated process. Each collector node is designed to efficiently collect data from specific segments of the IT environment. This includes devices, servers, network components, applications, databases, virtual machines, containers, and cloud resources, to name a few. By specializing in gathering data from specific sources, collector nodes optimize data acquisition and minimize the impact on the infrastructure.

Moreover, collector nodes employ a wide array of protocols and techniques to collect various types of data. These protocols can range from standard SNMP (Simple Network Management Protocol) for network devices to API-based methods for cloud resources, WMI (Windows Management Instrumentation) for Windows-based systems, and other customized methods for different application platforms.

Additionally, collector nodes have the capability to collect not only metrics but also events, logs, and other relevant information. Metrics provide real-time performance data, while events and logs offer insights into system activities, errors, and anomalies. The combination of diverse data types enriches SL1's data repository, allowing for a comprehensive and contextualized understanding of the IT environment's behavior.

The decentralized nature of collector nodes also enhances the scalability and resilience of SL1's monitoring architecture. By distributing the data collection load across multiple nodes, SL1 can effectively scale to monitor large and complex environments with thousands or even millions of monitored elements. Additionally, if a collector node becomes temporarily unavailable due to maintenance or network issues, other nodes can seamlessly continue data collection, ensuring uninterrupted monitoring.

Furthermore, collector nodes support secure communication between the monitored assets and the SL1 platform. Data integrity and confidentiality are paramount, and SL1's collector nodes employ encryption and secure communication protocols to protect sensitive information during transit.

The information collected by collector nodes serves as the foundation for SL1's analytics, automation, and visualization capabilities. By centralizing and processing this vast amount of data, SL1 can generate actionable insights, automate routine tasks, and create intuitive dashboards that provide IT teams with valuable information for decision-making and problem-solving.

Processing Nodes

Processing nodes play a pivotal role in transforming the vast amounts of collected data into meaningful and actionable insights. Once the data is transmitted from the collector nodes, the processing nodes take charge of the data processing, analysis, and correlation tasks. These nodes are designed to handle the data in a scalable and efficient manner, leveraging advanced algorithms, machine learning, and analytics techniques to extract valuable information from the raw data.

The sheer volume of data collected from diverse sources, including devices, servers, applications, and cloud resources, can be overwhelming for IT teams to manually process and interpret. The processing nodes act as the data processing powerhouse, capable of handling large-scale data streams in real time. By efficiently ingesting, parsing, and storing the data, these nodes ensure that nothing is lost in the process, and all relevant information is available for analysis.

Processing nodes harness the power of advanced algorithms and machine learning to analyze the data comprehensively. Machine learning models continuously learn from historical data and identify patterns, trends, and anomalies that may indicate potential issues or opportunities for optimization. These predictive capabilities enable SL1 to move beyond reactive monitoring and adopt a proactive approach, identifying potential problems before they escalate into critical incidents.

The correlation capabilities of the processing nodes are crucial in understanding the interconnectedness of data and events across the IT environment. By correlating data from various sources, the nodes can identify cause-and-effect relationships between seemingly unrelated events. This correlation is particularly valuable when troubleshooting complex issues that may involve multiple components and services. With this comprehensive view, IT teams can pinpoint the root cause of incidents more accurately and resolve them efficiently.

Moreover, processing nodes excel at contextualizing the data, providing a deeper understanding of its significance. Raw data streams may not always reveal the context or impact of specific events, making it challenging for IT teams to prioritize their response efforts. Through data enrichment and contextual analysis, the processing nodes add valuable metadata, timestamps, and other relevant information to the data, enabling better decision-making and incident prioritization.

The Processing Nodes generate actionable information for IT teams as they perform sophisticated data analysis. This information is presented through customizable dashboards, reports, and alerts, providing IT administrators with real-time insights into the health and performance of the IT environment. With this actionable information at their fingertips, IT teams can make informed decisions, proactively address potential issues, and optimize their IT infrastructure to meet business needs effectively.

Furthermore, the processing nodes support a wide range of analytical techniques to cater to different types of data and use cases. From statistical analysis for performance metrics to log pattern recognition for anomaly detection, SL1's processing nodes cover diverse analytical methods. This versatility ensures that the platform is well equipped to handle the complexity and diversity of modern IT environments.

Database Nodes

SL1's architecture incorporates database nodes that serve as the backbone for data storage and management. These nodes store the collected data as well as the processed and analyzed information. The database nodes use scalable and robust database technologies to ensure efficient storage, retrieval, and accessibility of data.

One of SL1's database nodes' key strengths lies in their ability to support high-performance queries. In an era where rapid decision-making and real-time insights are paramount, SL1's architecture ensures that users can execute intricate and resource-intensive queries without compromising on speed or efficiency. This capability is instrumental in empowering

IT professionals, allowing them to swiftly extract actionable insights from the wealth of data at their disposal.

Presentation Nodes

Presentation nodes handle the user interface and provide visual representations of data and information to users. These nodes deliver the dashboard views, reports, and analytics that IT teams interact with to monitor and manage the IT infrastructure effectively.

Presentation nodes are designed to provide users with not just a functional interface but also an intuitive and user-friendly one. Navigating through the vast sea of data generated by modern IT environments can be daunting, but SL1's presentation nodes aim to simplify this process. They offer a carefully crafted, ergonomic interface that guides users seamlessly through the platform, making it easy for both seasoned IT professionals and newcomers to access the information they need.

Customization is a cornerstone of SL1's presentation nodes. Recognizing that each organization has unique requirements and preferences, these nodes empower users with the ability to tailor their experience. Users can personalize dashboards, reports, and visualizations to align with their specific objectives and priorities. This adaptability ensures that SL1 caters to a wide range of use cases and industries, from healthcare to finance, from e-commerce to telecommunications.

Moreover, the intuitive nature of SL1's presentation nodes extends beyond aesthetics. They are designed to foster interaction with data and insights. Users can not only view data but also actively engage with it, drilling down into details, setting alerts, and triggering actions based on real-time information. This interactivity empowers organizations to proactively manage their IT operations, respond swiftly to anomalies, and optimize their infrastructure.

Furthermore, presentation nodes serve as a bridge between the complexity of IT data and the clarity of meaningful insights. They translate the technical jargon and intricate metrics into digestible information that decision-makers can readily comprehend. This is essential in fostering collaboration between IT teams and stakeholders from other departments who may not have a deep technical background.

Communication and Coordination

The nodes within the SL1 architecture communicate and coordinate with each other to ensure seamless operation. They exchange data, status updates, and commands to maintain synchronization and consistency across the system. This communication and coordination mechanism allows SL1 to efficiently distribute tasks and workload among the nodes, adapt to changing IT environments, and ensure the overall effectiveness and performance of the system.

The advantages of SL1's node-based architecture should be obvious. SL1 works like Nodeville. The design facilitates scalability and maximum availability by allowing additional nodes to join in and handle increased data volumes or to accommodate growing infrastructures.

It's not an exaggeration so say that SL1 can effectively monitor and manage even the most complex and dynamic environments.

6

Configuring SL1

You're an IT operations manager at a growing corporation. You've been asked to minimize downtime, streamline your team, and make work move faster. You read this book a few months ago and were sufficiently intrigued by the prospect of AIOps to help you add more value to your company's bottom line. You scheduled and completed a productive meeting with a team of ScienceLogic consultants, where a pathway to boosting your organization's observability capabilities became clear. You were able to effectively communicate this to your own key stakeholders, and they love the idea (especially the "less downtime" part).

With the green light to proceed, and ScienceLogic standing ready to get things moving, now what?

What are the concrete steps involved in configuring ScienceLogic's SL1 platform to meet your organization's specific needs and objectives?

The fact is that configuring SL1 isn't just about flipping a few switches or adjusting a couple of settings. Configuring SL1 is a nuanced process that requires careful planning and expertise to ensure it aligns perfectly with your IT infrastructure and operational goals.

In this chapter, we'll delve into the details of configuring SL1, exploring the various components and customization options available to you. This comprehensive guide will equip you with the knowledge and insights needed to harness the full potential of ScienceLogic's AIOps solution.

The first decision an organization has to make is whether to run SL1 entirely in its own cloud or to take advantage of ScienceLogic's SaaS offering.

SaaS

ScienceLogic's SaaS offering presents a transformative solution for businesses seeking seamless deployment of the powerful SL1 platform *without* the complexities of managing deployment architecture. With this innovative approach, companies can shift their focus from the intricacies of infrastructure management to optimizing their operational efficiency and business outcomes.

Software as a Service (SaaS) is a cloud computing model and software distribution approach in which applications are hosted and made available to users over the internet as services, often on a subscription basis. In the SaaS model, software applications are centrally managed and maintained by a third-party provider, eliminating the need for users to install, update, or manage the software on their local devices or servers. Here's a comprehensive definition of SaaS, along with some examples:

Key Characteristics of SaaS:

1. **Accessibility:** SaaS applications are accessible via web browsers, allowing users to access them from virtually any internet-enabled device, such as laptops, tablets, and smartphones.
2. **Subscription-Based:** SaaS is typically offered through subscription plans, where users pay recurring fees (e.g., monthly or annually) for access to the software. This pricing model often includes software updates, customer support, and maintenance.
3. **Multi-Tenancy:** SaaS providers serve multiple customers (tenants) from a completely or partially shared infrastructure, with each tenant's data securely separated to maintain data privacy and security.
4. **Automatic Updates:** SaaS providers handle software updates, ensuring users can access the latest features, bug fixes, and security enhancements without manual installations.
5. **Scalability:** SaaS applications are designed to scale quickly, accommodating changes in the number of users or resource needs, making them suitable for businesses of all sizes.
6. **Managed Security:** SaaS providers take responsibility for securing the software, infrastructure, and user data, implementing robust security measures to protect against cyber threats.
7. **Reduced IT Overhead:** Organizations using SaaS can reduce IT infrastructure costs and responsibilities as

the provider manages servers, databases, and software maintenance.

8. **Customization and Integration:** Many SaaS applications offer customization options and support integration with other SaaS tools or on-premises systems through APIs.

9. **Pay-as-You-Go:** SaaS often employs a pay-as-you-go model, where organizations are charged based on their actual usage, providing cost flexibility.

10. **Collaboration and Remote Work:** SaaS solutions often include collaboration features, facilitating remote work with tools for document sharing, communication, and project management.

By embracing ScienceLogic's SaaS offering, organizations can bypass traditional concerns and complexities associated with deployment architecture. They no longer need to grapple with hardware procurement, software installations, or infrastructure maintenance. Instead, ScienceLogic assumes the responsibility for the underlying technical framework, allowing businesses to harness the full potential of the SL1 platform swiftly.

This streamlined experience begins with a simplified onboarding process. Companies can swiftly onboard the SL1 platform without the overhead of configuring hardware or setting up intricate software components. Whether a business operates on premises or in a hybrid environment, ScienceLogic's SaaS offering adapts to fit seamlessly, ensuring the platform aligns with each organization's unique technological landscape.

Furthermore, ScienceLogic's SaaS offering liberates businesses from the intricacies of deployment and ensures scalability. As organizations expand and adapt, the platform effortlessly scales to accommodate evolving requirements, all without placing additional burden on IT teams to manage and oversee these changes. This scalability is underpinned by ScienceLogic's robust infrastructure, enabling businesses to focus on optimizing their operations instead of provisioning and managing resources.

On-Premises

For users who want to run SL1 entirely on their own data center or cloud, ScienceLogic has three options. It's ultimately the user's decision, but ScienceLogic works closely with users to build and implement the right solution.

1. **All-in-One:** In this configuration, a single node or appliance provides all the functions of SL1. The capacity of an all-in-one instance cannot be increased by adding additional nodes or appliances. This configuration is best for smaller deployments.

2. **Distributed:** In this configuration, the functions of SL1 are divided between multiple nodes or appliances. A distributed instance of SL1 can be as small as two nodes or machines or include numerous instances of each node or appliance. This configuration is best for production environments that monitor many devices or a large volume of data for each device.

3. **Extended:** This is an extension of a distributed instance. The extended configuration adds both a compute cluster

and a storage cluster. The compute cluster includes multiple compute nodes. The storage cluster includes multiple storage nodes. The extended configuration also adds a management node to install and update the compute cluster and storage cluster, and one or more load balancers to manage the workload of the compute cluster. This configuration provides scale and can take advantage of the SL1 agent to collect detailed data about devices and applications, ultimately enriching SL1's ability to diagnose problems and suggest improvements.

Below, we'll discuss these different configurations in more detail. Even if you're not planning to implement SL1 currently, understanding how these configurations and instances differ can shed more light on exactly how SL1 works.

All-in-One Configuration

The "all-in-one" (AIO) configuration of ScienceLogic SL1 refers to a deployment setup where all the necessary components and functionalities of SL1 are consolidated into a single server or virtual machine. SL1 operates as a self-contained unit in this configuration, encompassing all the required components, such as the data collector, processor, database, and presentation layers, within a single system.

The AIO configuration offers a simplified and streamlined deployment option suitable for smaller environments or organizations with limited resources. Instead of setting up multiple nodes and distributed components, SL1 can be installed and run on a single server, reducing the complexity and resource requirements associated with a distributed architecture.

In an AIO configuration, the SL1 server acts as a central hub that performs all the essential data collection, processing, analysis, storage, and presentation functions. It collects data from various sources within the IT environment, including devices, servers, applications, and cloud resources, and processes it locally on the server. The collected data is then stored in the server's database, which resides on the same system. Finally, the SL1 server provides the user interface, presenting the data, insights, and visualizations through its integrated presentation layer.

While the AIO configuration simplifies the initial setup and management, it does have some limitations compared to a distributed architecture. The consolidation of all SL1 components into a single system may impact scalability and performance, particularly in large or complex IT environments with high data volumes. In such cases, a distributed architecture with separate collector, processor, database, and presentation nodes would be more suitable to handle the workload efficiently.

However, the AIO configuration remains a viable option for smaller environments, proof-of-concept deployments, or situations in which simplicity and resource constraints take precedence over scalability and advanced capabilities. It offers an all-encompassing solution within a single server or virtual machine, allowing organizations to leverage the comprehensive monitoring and management functionalities of ScienceLogic SL1 without the complexity of a distributed architecture.

The following features are not supported by all-in-one appliances:

1. Using a SAN for storage
2. Disaster recovery
3. High availability for database servers
4 High availability for data collectors
5. Additional data collectors, message collectors, or administration portals

Distributed SL1 Systems Configuration

In the distributed SL1 systems configuration SL1, the architecture consists of two required layers—the *database layer* and
the *collection layer*—and an optional *interface layer*. Let's delve
into the specifics of each layer:

Database Layer

The database layer is a crucial component of the distributed
architecture. It comprises one or more database nodes responsible for storing and managing the collected data, as well as the
processed and analyzed information. These nodes use scalable
and robust database technologies to handle the large volumes
of data generated by the environment. Within the database
layer, data from various sources is consolidated and stored efficiently. This allows for fast and reliable data retrieval, supporting
reporting, analytics, and historical data analysis. The database
layer ensures data integrity, consistency, and accessibility across
the distributed SL1 system.

The database layer includes the following:

1. The database server
2, The administration portal
3. API access

These functions are divided among multiple nodes or appliances.

Collection Layer

The collection layer encompasses the collector nodes strategically deployed throughout the IT environment. These nodes are responsible for gathering data from diverse sources, such as devices, servers, applications, and cloud resources. Collector nodes employ specific protocols and agents to collect metrics, events, logs, and other relevant data. The collection layer performs the essential function of data acquisition, ensuring comprehensive data collection from all relevant sources. It collects data in real time or at predefined intervals and forwards it to the subsequent layers for further processing and analysis. The collector nodes are designed to handle various data collection methods and protocols to accommodate the diverse IT infrastructure landscape.

The collection layer includes the following:

1. Data collectors
2. Message collectors

As with the database layer, these functions are divided among multiple nodes or appliances. Collectors can also be grouped, with load balancing and high availability across the collector group.

Interface Layer (optional)

The interface layer is an optional component that provides the user interface, dashboards, reports, and visualizations for interacting with the SL1 system. It allows IT teams and users to access and make sense of the collected data and insights provided by SL1. The interface layer enables users to navigate, customize, and interact with the SL1 system. It offers a user-friendly interface that facilitates monitoring, troubleshooting, and management tasks. Users can visualize data, configure alerts, generate reports, and gain insights into the performance and health of their IT infrastructure. The interface layer enhances user experience, enabling efficient and intuitive interactions with the distributed SL1 system.

Configuration Summary

In a distributed SL1 systems configuration, the architecture consists of several specialized components, including the user interface, database server, data collector, and message collectors. These components collectively perform the essential functions of SL1. The distribution of these functions can vary depending on the size and complexity of the SL1 system.

In larger SL1 systems, a dedicated node or appliance is assigned to each function. This means that separate nodes or appliances are responsible for handling the user interface, database server, data collector, and message collectors. This distributed setup allows for efficient scalability, performance, and fault tolerance, as each component can be scaled independently based on the workload and requirements.

On the other hand, in smaller SL1 systems, some nodes or appliances may perform multiple functions. For example, a single node or appliance may handle both the user interface and the database server functions. This consolidated approach simplifies the deployment and management process for smaller environments or organizations with limited resources.

Additionally, the all-in-one appliance system represents a configuration in which a single SL1 node or appliance performs all four functions. In this setup, all the required components, including the user interface, database server, data collector, and message collectors, are consolidated into a single system. This configuration offers a simplified and streamlined deployment option that is particularly suitable for smaller environments where simplicity and resource constraints take precedence over scalability and advanced capabilities.

Whether in a distributed system or an all-in-one appliance configuration, the SL1 components work together to provide comprehensive monitoring and management capabilities. The user interface enables users to interact with the system, while the database server stores and manages the collected data. The data collector gathers data from various sources, and the message collectors handle the processing and storage of messages and events.

These components collaborate to deliver efficient and effective monitoring and management of the IT infrastructure.

7

Onboarding SL1

Analytics and Dashboards

IN SL1, A dashboard is a page that displays one or more graphical reports called widgets. Each widget appears in its own pane and displays charts, tables, and text. Access to dashboards is limited and is based on a user's login credentials. Also, some dashboards might be private instead of public, so they are available only to certain specified users on a case-by-case basis.

SL1 includes several system default dashboards, but you can also create your own dashboards that are completely customized to your specific data visualization needs. When you create a dashboard, you are defining a container that will display widgets. Each widget in a dashboard displays a report about data in SL1. You must define a name for the dashboard, specify the space for one or more widgets, and determine the settings for those widgets.

SL1 Default Dashboards

1. **NOC Overview:** This dashboard displays a high-level overview of your business services and their current health statuses in a single-pane view. This dashboard is

99

updated in real time to reflect the most recent information about each service.

2. **Business Services Dashboard:** This dashboard displays availability, health, and risk data for all of your business services.

3. **Business Service Details Dashboard:** This dashboard provides an overview of the services that belong to a specific business service.

4. **IT Services Dashboard:** This displays availability, health, and risk data for all of your IT services.

5. **IT Services Details Dashboard:** This provides an overview of the device services that belong to a specific IT service.

6. **Device Services Dashboard:** This dashboard displays availability, health, and risk data for all of your business devices.

7. **Device Service Details Dashboard:** This dashboard provides an overview of the devices that belong to a specific device service.

8. **Server Dashboard:** This dashboard displays CPU usage, memory usage, disk usage, and other widgets related to servers. It includes widgets on server leaderboard, CPU usage, memory usage, disk usage, swap, total network traffic, latency (the amount of time it takes SL1 to communicate with the device), and availability.

In SL1, you first select from a list of pre-defined widgets, and then you customize what will be displayed by supplying values in the option fields provided by each widget. Your widgets will update in real time as more data becomes available.

As with many other applications, SL1 allows you to select one or more dashboards so that they always display at the top of the list on your dashboards page. This process is called "favoriting dashboards" or "favorite dashboard," and it's a quick way for administrators and analysts to get a bird's-eye view of what's most important to them.

A leaderboard widget allows a dashboard user to select specific items in a widget so that data about only those items displays in other widgets in the dashboard. In SL1, this feature is called "driving data" or "driving the context" of a dashboard widget.

For example, in the server leaderboard widget pictured above, if you select one or more servers on the leaderboard widget, the other widgets in the dashboard will display data about just the servers you selected. The other widgets receive the context from the "driving" widget, which in this example is the leaderboard widget.

SL1 dashboards also feature a *time span* filter, where you can adjust the time span of data that appears in all the widgets on your dashboard. By default, SL1 uses a time span of the *last 6 hours*, but you can update the time span of data reflected in a dashboard based on your needs.

SNOW Integration

It's become an iron law of business: Data lies at the core of modern enterprises, holding more potential to revolutionize operations and drive business success than any other asset.

This means, of course, that poor data quality can prove disastrous, resulting in slowed response times, operational inefficiencies, and a frustrating user and customer experience. As enterprises scale and data fragments, the risk of inaccurate data looms large, necessitating a robust solution to ensure data accuracy and reliability.

ScienceLogic SL1 + ServiceNow

Integrating SL1 with ServiceNow presents a game-changing synergy that addresses the challenges of data accuracy and management head-on. SL1's real-time, context-rich data feeds directly into the configuration management database (CMDB), eliminating the need for manual reconciliation and population. As a result, IT service management processes, such as incident, problem, and change management, receive an unprecedented boost in automation, streamlining operations and freeing resources for strategic initiatives.

ServiceNow is a leading cloud-based platform that revolutionizes the way enterprises manage and optimize their digital workflows and IT services. With a relentless focus on enhancing organizational productivity and delivering exceptional customer experiences, ServiceNow offers a comprehensive suite of applications and solutions that streamline complex business processes, drive efficiency, and foster collaboration across departments. At its core, ServiceNow serves as a transformative force, replacing fragmented and siloed systems with a unified and intelligent approach to managing tasks, incidents, problems, changes, and service requests.

From IT service management (ITSM) to human resources, customer service, security operations, and beyond, ServiceNow's powerful platform empowers organizations to automate and orchestrate a vast array of operations, elevating employee productivity and enabling faster decision-making. Its robust capabilities extend beyond IT, bridging the gap between different business functions and propelling them toward a more agile and responsive future. With cloud-based architecture, extensive analytics, and machine learning capabilities, ServiceNow transcends traditional boundaries, allowing organizations to embrace the true essence of digital transformation and embark on a journey of continuous innovation and growth.

Accelerating Problem Resolution

A key benefit of the ScienceLogic-ServiceNow integration lies in its ability to enable IT operations to identify and resolve problems impacting business services with remarkable speed and efficiency. By furnishing Ops and DevOps teams with better problem management data, SL1 empowers these teams to act proactively and respond swiftly to incidents, mitigating their impact on critical applications and services.

The outcome is not only more resilient digital experiences but also heightened customer and employee satisfaction.

Unlocking the Power of Automation

The SL1-ServiceNow partnership sets the stage for ground-breaking automation capabilities. The linchpin here is the unwavering commitment to maintaining real-time CMDB accuracy, which acts as the cornerstone for driving intelligent data and workflow automations.

Through this collaborative approach, IT operations finds itself uniquely positioned to transcend the limitations of manual intervention and systematically streamline repetitive tasks, all while optimizing resource allocation. This deeply ingrained automation-centric approach serves as a catalyst for organizations seeking not just incremental improvements but also a quantum leap in operational efficiency. As the wheels of automation turn, the organization's capacity to maximize productivity surges, with IT teams liberated to focus on strategic initiatives and innovation. The ultimate result is a holistic elevation of service levels, delivering a seamless and consistently superior experience to customers and end-users, reinforcing the organization's competitive edge and future-proofing its operations in a rapidly evolving digital landscape.

Accelerating Mission-Critical Efforts

For organizations relying on mission-critical applications and services, the ScienceLogic–ServiceNow integration serves as an invaluable ally. By reducing incident noise and providing real-time insights, SL1 expedites break/fix efforts, minimizing downtime and bolstering the resilience of essential applications.

As IT organizations face mounting pressures to deliver exceptional digital experiences, the seamless integration of SL1 with ServiceNow emerges as a powerful solution. This synergy provides accurate and context-rich data that fuels automation, streamlines IT service management processes, and accelerates problem resolution. By unlocking the true potential of data and automation, ScienceLogic and ServiceNow pave the way for organizations to thrive in the face of complexity, setting new benchmarks for operational efficiency and customer satisfaction. The journey to a transformative future, where technology meets seamless synergy, has just begun—welcome to a realm of unparalleled integration and success.

8

Crawl, Walk, Run

"A GREAT EXPERIENCE for our people, a great experience for our customers, having a service mix that's relevant for our market, a brand with a purpose, and executing really well."

That's what's most important to Datacom, according to Justin Gray—Managing Director of Datacom New Zealand.

As a wholly New Zealand-owned IT services company, Datacom has a long history in the IT industry. Founded in 1965, it's steadily grown to become one of the largest IT services providers in the Australasian region—a household name, really, in the IT services space "Down Under."

These days, Datacom offers a wide range of IT services and solutions to businesses, government agencies, and organizations across various industries. These services encompass areas such as software development, IT infrastructure services, managed IT services, digital transformation, CRM, Business Process Outsourcing, and IT consulting services.

Their success is built on a commitment to innovation, customer service, and delivering high-quality IT solutions. They work with a diverse range of clients, including government agencies, healthcare providers, financial institutions, and commercial enterprises, helping them harness the power of technology to achieve their goals.

But in 2020, Datacom came to ScienceLogic with some pretty big ideas.

"Our customers started demanding something different," says Alexandra Coates, Managing Director. "They wanted contemporary offerings backed by automation."

No longer were incremental improvements or traditional IT solutions enough. This shift in customer preferences reflected the broader industry trends that had been gaining momentum. Traditional IT solutions had become insufficient in the face of rapidly advancing technologies and increasing business complexities. Customers were no longer satisfied with gradual enhancements or manual interventions. They were looking for innovative, forward-thinking solutions that could keep pace with the rapid changes in the digital landscape.

A tall order. But Datacom saw the value in being first to the AIOps table. They expressed as much to ScienceLogic's engineers, who were as eager as Datacom to build a comprehensive and future-proof solution that future-proofed Datacom's IT services offerings.

By choosing a single product, ScienceLogic SL1, as their central point of focus, Datacom aimed to consolidate and unify their IT operations. This centralization allowed them to eliminate silos that often exist when managing IT separately for different customers or departments. Having a common platform enabled them to standardize processes, reduce complexity, and improve overall efficiency. And the concept of a "central brain" represented an intelligent and data-driven system that could collect, analyze, and interpret vast amounts of information from various sources. This centralized intelligence would serve as the backbone for making informed decisions and driving operational improvements. It acted as the central hub where insights from different customer environments were processed and turned into actionable information.

One of the key advantages of this approach would be Datacom's ability to gain insights across all of their customers. Traditional IT management too often results in fragmented data, making it challenging to identify overarching trends or issues. With a central capability and a "central brain," Datacom could aggregate data from all their clients, providing a comprehensive view of the IT landscape.

But of course, the ultimate goal of this transformation was to create value—not just for Datacom but also for their clients. By centralizing operations and using automation, Datacom could proactively identify opportunities for improvement, troubleshoot issues more effectively, and enhance the overall performance of IT systems. This added value translated into better service delivery and improved outcomes for Datacom's customers.

"This is ultimately about changing the way that our 2,000 people who deliver IT-managed services think, make decisions, and work every day," says Gray.

No small task.

In an era where innovation and efficiency reign supreme, Datacom and ScienceLogic came together with a singular focus: to apply cutting-edge technologies and methodologies to usher enterprises into a new era of IT operations. Their shared vision was to propel organizations toward an automated state of operations, where manual interventions would be minimized, and agility maximized.

Formalizing their partnership involved more than just handshakes and agreements; it was underpinned by mutual investments and a commitment to deliver tangible results:

A Phased Approach: Crawl, Walk, and Run

One of the cornerstones of this strategic collaboration was ScienceLogic's customer success team working hand-in-hand with Datacom. Together, they developed a comprehensive benefits realization plan. This plan was not just theoretical; it was brought to life through customer workshops that meticulously outlined a phased approach to projects. Datacom and ScienceLogic recognized that transformation was not a sprint, but rather a journey comprising three distinct phases: crawl, walk, and run.

In the "Crawl" phase, the focus was on establishing a baseline understanding of the existing IT infrastructure. Datacom and ScienceLogic conducted thorough assessments, identifying pain points, and laying the foundation for subsequent improvements.

The "Walk" phase was characterized by gradual enhancements. Here, Datacom began implementing ScienceLogic's AIOps platform, SL1, in a controlled manner. They addressed critical issues, optimized processes, and honed their automation capabilities.

Finally, in the "Run" phase, Datacom and ScienceLogic unleashed the full potential of automation. Routine tasks were automated, allowing IT teams to redirect their efforts toward more strategic initiatives. The transformation was now palpable, with increased efficiency and reduced downtime.

At the core of Datacom's market offer was ScienceLogic's AIOps platform, affectionately known as SL1. This platform became the bedrock upon which Datacom's digital transformation solutions were built. It was the bridge between legacy systems and the dynamic, automated future that both companies envisioned.

SL1 provided Datacom with real-time monitoring capabilities, offering insights into network performance, server health, and application performance. This robust foundation empowered Datacom to proactively identify and address issues, ensuring the reliability of their IT systems.

In their commitment to delivering results, Datacom and ScienceLogic established a success-based commercial construct. This framework was carefully designed to ensure that the right incentives were in place to drive the desired business outcomes. Success was not just measured in terms of product adoption, but also through the lens of customer satisfaction, cementing a customer-centric approach.

This alignment of incentives created a win-win scenario. Datacom was motivated to excel in implementing ScienceLogic's solutions, while ScienceLogic was equally committed to providing top-notch support and guidance.

Datacom tapped into ScienceLogic's JumpStart program to assist in the evolution of their service catalog. This proved invaluable in planning the introduction of new services and evolving existing ones. The collaborative approach of Datacom and ScienceLogic ensured that the service catalog was always aligned with the ever-changing demands of the digital landscape.

The JumpStart program was not a mere toolkit; it was a strategic partnership in itself. Datacom's service offerings were enhanced and expanded, providing their clients with cutting-edge solutions that met their evolving needs.

When it came to strategic accounts, Datacom and ScienceLogic were united. They provided bid support through a combination of compelling content presentations, informative collateral, and knowledgeable personnel. This collaborative approach strengthened their position in the market and showcased the depth of their partnership.

In these strategic engagements, Datacom and ScienceLogic were more than vendors; they were trusted advisors. Their joint expertise helped Datacom's clients navigate complex IT landscapes, offering tailored solutions that addressed specific challenges and objectives.

The commitment to excellence didn't stop with the deployment of solutions; it extended into operational support. Datacom leveraged ScienceLogic's remote admin service to ensure best-in-class support for the SL1 platform. This wasn't just about resolving issues but also about knowledge transfer, empowering Datacom's teams to excel in their operations.

The remote admin service was a testament to ScienceLogic's dedication to customer success. It was not just a helpdesk but a collaborative partnership, where Datacom and ScienceLogic worked together to ensure the SL1 platform operated at peak efficiency.

In the partnership between Datacom and ScienceLogic, the transformation of IT operations wasn't just a goal; it was a shared vision. Together, they harnessed the power of automation, cutting-edge technology, and a customer-centric approach to steer their clients toward a future where efficiency, innovation, and success converged. This was more than a partnership; it was a dynamic alliance propelling businesses into the digital age. It was a testament to the power of collaboration in achieving digital excellence.

The result?

"When you talk to the industry analysts, they would say that we're one of very few organizations tackling AIOps in a truly delivered way," says Coates.

9

SL1 Stack Buildout

Accounts/SSO/Permissions

LIKE ANY MODERN solution worth its chops, SL1 administrators can log in to their SL1 platform via single sign-on (SSO).

To implement SSO with SL1, organizations typically follow a series of steps. They begin by selecting an identity provider (IdP), a system responsible for managing user identities and authentication. Popular choices for IdPs include Microsoft Azure Active Directory, Okta, Ping Identity, or even a custom IdP designed specifically for the organization's needs.

Once you've chosen an IdP, the next step involves configuring ScienceLogic SL1 itself. This is done through SL1's administrative interface, in which users navigate to the security or authentication settings. Within the SL1 configuration, administrators define the specifics of the SSO setup. This includes entering details such as the IdP's metadata URL or certificate and configuring how user attributes from the IdP correspond to ScienceLogic SL1 accounts.

To ensure a smooth integration, it's important to conduct a thorough test of the SSO setup. This test typically involves performing a trial SSO login to confirm that all configurations are functioning correctly. Any issues or discrepancies can be identified and addressed before deploying SSO for all users.

With successful testing, organizations can proceed to enable SSO for all users who should benefit from this streamlined authentication method.

The user experience with SSO involves a seamless login process. When a user attempts to log in, he or she is redirected to the IdP's login page. There, the user enters the credentials (username and password) as usual. Upon successful authentication, the IdP generates a token or assertion, signaling the user's verified identity.

ScienceLogic SL1 then processes this token or assertion, confirming the user's identity and granting access to the platform. The user is seamlessly logged in to ScienceLogic SL1 without the need for a separate username and password for the platform.

Ongoing management of SSO users and their permissions is handled through the IdP. Any changes made in the IdP, such as adding or removing users, automatically apply to ScienceLogic SL1, simplifying user management.

Regular monitoring of SSO access to SL1 is important to ensure continued smooth operation. It's also essential to keep both the IdP and ScienceLogic SL1 configurations up-to-date to maintain the integration effectively. This ensures that users experience the benefits of SSO while maintaining the security and efficiency of the authentication process.

Powerpacks / Dynamic Apps— which to include or exclude

ScienceLogic SL1 PowerPacks refer to pre-configured monitoring templates or modules designed to extend the functionality of ScienceLogic SL1, particularly in the context of IT infrastructure and application monitoring. PowerPacks are designed to simplify the process of setting up and configuring monitoring for specific technologies, devices, or applications.

Here are some key points about ScienceLogic SL1 PowerPacks:

1. **Monitoring Templates:** PowerPacks are essentially monitoring templates that are tailored for specific technology stacks or components commonly found in IT environments. They provide a set of predefined monitoring configurations, rules, and data collection methods.

2. **Ease of Deployment:** PowerPacks make it easier for IT teams to deploy monitoring for various technologies without having to manually configure each monitoring aspect. This can save time and reduce the chance of configuration errors.

3. **Coverage:** They cover a wide range of technologies, including servers, storage devices, network equipment, databases, cloud services, and more. This broad coverage allows organizations to monitor their entire IT ecosystem comprehensively.

4. **Customization:** While PowerPacks offer predefined monitoring configurations, they are typically customizable.

Users can tailor the settings to match their specific requirements and infrastructure.

5. **Updates and Maintenance:** ScienceLogic updates and maintains the PowerPacks to ensure that they stay current with changes in technology and best practices. This helps organizations keep their monitoring configurations up-to-date.

6. **Integration:** PowerPacks can often be integrated with other IT management tools and services, allowing for a seamless monitoring experience across the entire IT stack.

7. **Community Contributions:** In some cases, ScienceLogic SL1 users and the broader IT community may contribute their own PowerPacks to address specific monitoring needs that are not covered by the out-of-the-box offerings.

Organizations and Device Groups

In SL1, organizations and device groups are organizational and structural components used to manage and organize the monitoring of IT resources. SL1 uses organizations and device groups to provide a structured approach to managing and monitoring IT resources in complex environments. Organizations help in dividing responsibilities and tailoring monitoring configurations to specific units or departments, while device groups enable the finer-grained organization of devices and applications based on common attributes or criteria. This organizational structure enhances the efficiency and effectiveness of IT monitoring and management with SL1.

Here's an explanation of each:

1. **Organizations:** In SL1, organizations represent distinct units or entities within your overall IT environment. These units can be defined based on different criteria such as departments, business units, geographical locations, or any other logical divisions that make sense for your organization's structure.

   Organizations allow you to segment and manage the monitoring of IT resources separately for each unit. Each organization can have its own set of monitoring configurations, access controls, and dashboards tailored to its specific needs.

   This organizational structure is particularly useful for large enterprises with diverse IT environments, as it helps in delegating responsibilities and providing focused visibility and control to different teams or units.

2. **Device Groups:** Device groups, on the other hand, are used to further categorize and organize specific sets of IT resources within an organization. These resources can include servers, network devices, applications, and more.

   Device groups help you group similar devices or components together for easier management and monitoring. For example, you might create device groups based on technology stacks (e.g., web servers, database servers), locations (e.g., data centers, remote offices), or other criteria relevant to your monitoring needs.

   By organizing resources into device groups, you can apply monitoring policies, templates, and rules more efficiently. It allows you to target specific configurations to different sets of devices based on their common characteristics.

Device groups also help in troubleshooting and incident management. When an issue arises, you can quickly identify and isolate the affected resources by looking at their respective device groups.

Discover: Bringing Devices into SL1 for Monitoring

Bringing devices into SL1 for monitoring is crucial for maintaining a healthy and efficient IT environment. It's what makes SL1 work.

Organizations can bring devices into SL1 for monitoring by following a structured process that involves discovery, configuration, and integration.

First is **device discovery.** With **auto-discovery,** SL1 will automatically identify devices within your IT environment. This process involves scanning your network for devices and collecting information about them. Auto-discovery is a convenient way to identify and add devices quickly.

Manual discovery is also possible for devices that may not be automatically discovered, or simply in cases where you want more control. To do this, you'll need to provide specific details about each device, such as its IP address, hostname, or other relevant information.

Once devices are discovered or manually added, you'll need to configure monitoring settings for each device. This involves selecting the appropriate monitoring templates or settings that match the type of device you're adding. SL1 typically includes a library of pre-configured monitoring templates for various device types, making it easier to apply the right settings.

Customize the monitoring configuration as needed to align with your organization's specific monitoring requirements. This may include setting performance thresholds, defining alerting rules, and specifying data collection intervals.

Depending on the type of device and the monitoring protocols used, you may need to provide credentials (e.g., usernames and passwords) to SL1 for secure access to the devices. This is particularly important for such devices as servers, routers, and switches. Ensure that the credentials you provide have the necessary permissions to collect monitoring data from the devices.

It should go without saying that SL1 is often designed to integrate with a variety of network management protocols and technologies. Ensure that your network and devices support the protocols required for SL1 to communicate with and monitor your devices. Common protocols include SNMP (Simple Network Management Protocol), WMI (Windows Management Instrumentation), SSH (Secure Shell), and more. Set up the necessary integration parameters, such as SNMP community strings or WMI credentials, to establish communication between SL1 and your devices.

Once you've determined that the system is working properly, you can deploy the monitoring configurations to the devices in your production environment. This process may involve applying monitoring templates, verifying settings, and activating monitoring for each device. Continue to manage your monitored devices within SL1 as your IT environment evolves.

This includes updating configurations, adding new devices, and removing devices that are no longer in use.

Agent vs. Agentless

SL1 offers two primary methods for monitoring IT resources: **agent-based** and **agentless** monitoring. These two approaches have distinct characteristics and are chosen based on specific requirements and use cases. Let's explore the differences between agent and agentless SL1 monitoring.

In **agent-based monitoring**, small software components known as "agents" are installed on the target devices or servers that you want to monitor. These agents collect data locally from the monitored system, which often includes performance metrics, system health information, and application-specific data. This local data collection can provide highly granular insights into the resource's behavior.

1. **Real-Time Monitoring:** Agent-based monitoring typically provides real-time data because agents continuously collect and transmit information to the central ScienceLogic SL1 platform.
2. **Security Considerations:** Since agents reside on the monitored devices, they require proper security considerations. Organizations must ensure that agents are securely deployed and that data transmission from the agents to the central platform is encrypted and protected.
3. **Resource Overhead:** Agents consume some system resources (CPU and memory) on the monitored devices.

While modern agents are designed to have minimal impact, it's important to assess the resource overhead, especially on critical systems.

Agentless monitoring, as the name suggests, doesn't require the installation of agents on the monitored devices. Instead, it relies on existing protocols and APIs to gather data remotely. It collects data from the target devices without any software footprint on those devices. It typically uses protocols like SNMP (Simple Network Management Protocol), WMI (Windows Management Instrumentation), or SSH (Secure Shell) for data retrieval.

1. **Ease of Deployment:** Agentless monitoring is often easier and quicker to set up since it doesn't involve deploying and managing software agents on each monitored device.
2. **Reduced Resource Impact:** Since no agents are running the monitored devices, there is no resource overhead on those devices. This makes agentless monitoring suitable for systems where resource consumption must be minimal.
3 **Protocol Compatibility:** Agentless monitoring's effectiveness depends on the availability and compatibility of the chosen monitoring protocols. Some systems may not support the necessary protocols for agentless monitoring.

How should users consider which of these two approaches to use with their SL1 setup? This depends on several criteria.

If minimizing resource consumption on monitored devices is crucial, agentless monitoring is preferred. Agent-based monitoring excels in highly detailed and granular monitoring by collecting a broader range of data from local systems. Compatibility with the monitoring protocols and technologies supported by target devices also plays a role in the decision, as some systems may necessitate agent-based monitoring when suitable protocols for agentless monitoring are unavailable. Security considerations also come into play, with agent-based monitoring offering greater control over data transmission, making it a choice in environments with stricter security requirements.

Templating

SL1 templating is designed to streamline the setup and management of IT resource monitoring.

Templates in SL1 are predefined configurations or models that simplify the process of monitoring various devices, applications, or services. Think of your car's dashboard, but with hundreds of possible configurations of gauges and metrics.

These templates promote consistency and efficiency by allowing organizations to establish standard monitoring settings for specific device types or applications, reducing the risk of configuration errors. Moreover, they promote reusability, enabling users to create monitoring configurations once and apply them to multiple instances of the same resource. Templates are typically customizable to accommodate specific requirements while benefiting from built-in monitoring expertise. They expedite monitoring deployment, as users can apply templates rather than configuring each resource from scratch.

SL1 typically includes a library of pre-built templates for a wide range of devices and technologies, serving as valuable starting points for monitoring needs. These templates are based on the ScienceLogic team's years of experience working with SL1 users to optimize dashboards for various organizations across many verticals. These templates receive (optional) global updates to stay current with industry changes, ensuring the ongoing relevance and effectiveness of monitoring configurations. Users can also associate templates with device groups to apply consistent monitoring settings across similar devices.

Additionally, templates may offer other advanced features, such as threshold settings. With threshold settings, administrators can establish predefined limits for metrics such as CPU usage, memory utilization, network latency, and more. When these thresholds are breached, SL1 generates alerts or triggers automated actions, enabling proactive problem resolution. Fine-tuning these thresholds ensures that alerts are triggered when performance issues are genuinely critical, reducing unnecessary noise in the monitoring system. event correlation rules, and performance data collection intervals to fine-tune monitoring according to specific use cases.

Another of these more advanced features is the event correlation template, which can become quite sophisticated. Event correlation templates are used to identify meaningful patterns or relationships within the monitored data. Templates in SL1 may include pre-configured event correlation rules, which help in identifying complex issues that may not be apparent through simple threshold-based monitoring. For example, event correlation can help identify the root cause of a performance problem by analyzing the sequence of events leading up to it. These rules can be highly customizable, allowing organizations to adapt them to their specific use cases and requirements.

Performance data collection intervals are another. Effective monitoring involves collecting performance data at regular intervals to track changes and trends over time. Templates often include predefined data collection intervals that suit common use cases. For example, critical systems may require real-time or near-real-time data collection, while less critical resources may have longer collection intervals to conserve resources. Fine-tuning these intervals ensures that monitoring is both responsive and resource-efficient, aligning with the organization's priorities.

With these advanced SL1 templating features, organizations can tailor their monitoring configurations to align precisely with their unique needs and use cases. This level of customization and granularity enhances the accuracy of issue detection and contributes to more efficient resource allocation, improved capacity planning, and ultimately, a more robust and responsive IT environment.

10

How to Change Your Operational Process

THE INTEGRATION OF AIOps has revolutionized the way large IT organizations approach operational change, highlighting its significance in boosting efficiency and driving superior business value for users. AIOps leverages advanced machine learning and data analytics to enhance decision-making, automate tasks, and provide real-time insights into the IT environment. By embracing operational change in conjunction with AIOps, large IT organizations can achieve a new level of agility and responsiveness that directly translates into improved user experiences and enhanced business outcomes.

AIOps empowers IT organizations to proactively identify and address potential issues before they impact users, mitigating downtime and disruptions. Operational changes, when informed by AIOps insights, can be implemented with a deeper understanding of the system's current state and potential risks. This not only minimizes the chances of service interruptions but also streamlines change management processes. By using AIOps to predict and manage operational changes, IT organizations can optimize resource allocation, allocate maintenance efforts more effectively, and maintain a higher level of system availability, contributing to improved user satisfaction.

Furthermore, AIOps enables IT organizations to uncover valuable patterns and trends hidden within large datasets, offering a comprehensive view of system performance, user behavior, and emerging opportunities. This data-driven approach enhances the decision-making process when implementing operational changes. For instance, AIOps can identify areas of the IT infrastructure that are underutilized, enabling organizations to make informed decisions about resource allocation and capacity planning. This optimization leads to cost savings, better resource utilization, and ultimately, a more efficient IT environment that directly contributes to delivering enhanced business value to users.

The synergy between AIOps and operational change presents a transformative opportunity for large IT organizations to elevate their efficiency and the value they bring to users. By leveraging AI-driven insights, these organizations can make informed decisions, proactively manage potential disruptions, and optimize their systems to meet user expectations. The marriage of AIOps and operational change sets the stage for an IT landscape that is not only responsive and adaptable but is also a key driver of competitive advantage and user satisfaction in today's digital era.

Challenges in Today's Complex IT Environments

A good customer experience is one of the most important metrics of success for your enterprise.

To deliver information, transactions, and interactions quickly and efficiently to your customers, you need to rely on a vast collection of interconnected technologies that work seamlessly together. But as transactions grow in complexity and end users' expectations increase exponentially, so does your IT infrastructure.

Your IT backbone must be able to support all your company's essential business services. But if you are like most enterprises, your IT environment is composed of a hybrid, complex, and highly interrelated mix of technologies, including apps, networks, and platforms deployed in a wide variety of environments, including data centers and private and public clouds.

Your infrastructure must also be able to support the entire spectrum of user types, including mobile workers, contractors, partners, and customers. And everything must work seamlessly on any number of end-user devices, including laptops, mobile phones, and wearables accessed from different kinds of networks and in different locations around the world.

But just because your IT is complex doesn't mean your monitoring should be.

Each technology layer emits volumes of data that contain the information required to monitor, troubleshoot, and ultimately improve those experiences. With all this growing complexity and the large volume, velocity, and variety of big data, it is time to rethink your infrastructure monitoring.

Here are some facts:

1. According to Gartner, the average cost of IT downtime is $5,600 every minute. Ninety percent of organizations report that a single hour of downtime costs over $100,000.

2. According to a survey from Uptime Institute, 31 percent of responding organizations experienced a downtime incident or severe degradation in the last year, and 48 percent reported at least one outage at their site or at a service provider in the last three years.

The best way to ensure that IT issues are resolved quickly—or prevented altogether—is to monitor and troubleshoot the underlying infrastructure within a service context.

While observing any one element of the infrastructure stack is a straightforward proposition (since there are plenty of tools available for monitoring individual pieces of the puzzle), observing each piece within the context of a service in a single platform enables you to address the challenges of modern IT management.

Modern Monitoring 101

Hybrid infrastructure is the collective grouping of interconnected technologies that includes servers, routers, and storage arrays in addition to software-defined anything running in your data centers and clouds.

Infrastructure monitoring must provide visibility into all these technologies to actively diagnose performance, utilization, capacity, and bandwidth problems across the entire IT estate before an outage hinders customer experience.

Modern monitoring tools provide you with the ability to see what is happening across your organization's infrastructure to help teams prevent outages—alerting your team to potential downtime, resource saturation, and business impact.

When you have comprehensive visibility into your entire enterprise, the potential problems that infrastructure monitoring tools identify can be solved quickly and effectively.

Step 1: Map Relationships between the Infrastructure and Applications

Most of today's enterprises have significant gaps and overlaps in visibility across the datacenter, cloud, hybrid cloud, and containers, resulting in increased operational costs and the uncertainty of conflicting sources of truth.

You need to see across your entire IT ecosystem (what you have, what it is, and how it works together) and understand how each component is related to each other (topology). Organizations benefit from monitoring solutions that go beyond the infrastructure and show relationships through app-to-infrastructure mapping so that you can understand how the infrastructure and apps work together for a particular technology.

Every application—from a complex financial services application to something as common as email—has a lot of moving parts. For example, if performance is slow, dependency mapping lets you know where to look to understand where the bottleneck is, what resources might be overtaxed, and how to resolve the problem. Once you have these insights, you know your next course of action should be to move workloads around or add capacity to handle spikes in usage.

Also, applications today are not running on dedicated infrastructure. They are relying on different components that are spread out throughout your entire IT ecosystem and can be shared with other applications. This makes having visibility across all your applications so important because it enables you to pull resources from other components to support others.

An example of this would be an eCommerce website needing to allocate more IT resources on Black Friday and Cyber Monday.

Step 2: Compose Business Services

In a rapidly changing IT world with microservice-based, containerized workloads, monitoring at the device level is no longer practical. Forward-looking enterprises are shifting from tracking individual IT devices and apps to a single business services view across a heterogeneous mix of clouds.

To quickly resolve and remediate issues unique to your business, you must first understand your business from an IT perspective.

How do you do this? Begin by defining your business services. Because understanding which services are critical and which SLAs you are expected to meet helps you to know what you need to work on first so you can remediate and eventually automate. We'll get to automation in our next chapter.

What Is a Business Service?

A business service is one or more technical services that give value to both internal and external customers. Business services should align IT assets with the needs of a company's employees and customers, and they should support business goals, facilitating the ability of the company to be profitable.

Usually, a business service includes an associated Service Level Agreement (SLA) that specifies the terms of the service.

Examples of Business Services Include:

1. Verification of internet access or website hosting
2. Remote backups and remote storage
3 Payroll, online trading, online banking

There are many advantages to seeing across your entire hybrid IT ecosystem with a business service view. First, your IT devices are constantly in flux—from the software updates for your applications to the number of devices using your network at any given time. But your business services are a lot more stable and long-term, changing very rarely when compared to the dynamic complexities of IT components.

Having a long-term, service-centric view across your data centers, clouds, applications, and devices helps you understand the impact of IT on the business. A service-centric approach to IT combines different applications together into a service your business is selling to the end user. In this way, you can start managing/prioritizing your work, giving you the stability to understand and learn the behavior over time.

Another advantage of having a unified, service-centric view across your IT estate is having the ability to pull in business KPIs to measure IT's impact on business outcomes. Why is this important? Let's use online trading as an example. A financial institution wants to achieve a certain number of trades a day. If they are monitoring their IT at the server or application level, the business is unable to see the high-level impact of those trades on revenue.

A service-centric approach takes away all the minutia and noise, giving you a clear understanding of how your infrastructure is impacting the service so you can rapidly remediate before the end-user is impacted.

If or when something goes wrong anywhere in your infrastructure—which is often shared across multiple applications and business services—you can quickly identify the impact and isolate the root cause.

Step 3: Add Analytics and Automation

While infrastructure monitoring is essential in knowing what you have and what it is (discovery) and in accurately assessing the health, availability, and risk across your complex ecosystem, forward-looking companies are now looking to automate, eliminating manual operations so they can respond proactively and reactively.

Now that you have composed business services and know the most common issues affecting your most critical services, you are ready to harness the power of machine learning to gain actionable insights from your infrastructure data.

Machine learning can detect weird or anomalous service behavior and can correlate those anomalies and common events within a service context. With machine learning, you can cut through the noise to quickly establish the root cause of an issue, enabling your team to keep in front of constantly changing environments.

Adding machine learning to an infrastructure monitoring tool can unlock powerful opportunities for the ITOps team.

Developing learning patterns of behavior on the service rather than the individual components allows you to automate the right things. What are the "right things" you should automate? Let's start with automated ticketing and routing. Industry averages indicate manual ticket creation and routing routinely takes more than 60 minutes. You can avoid a costly service impact and downtime by eliminating these time-consuming manual activities.

You can also automate troubleshooting and remediation steps. By automating entire workflows, such as the steps required to remotely log in to a device, gather diagnostics, and diagnose or remediate a problem, you can automatically capture diagnostic data to enrich both events and incidents. This enables faster root cause analysis and improved mean time to repair (MTTR).

If you are automating data collection, you can improve problem management, and then you will be better informed about what you can resolve. And once you know the commonality in that data—through mapping and monitoring by the service (not the device)—you can automate self-healing and resolution of issues.

When repetitive tasks and processes are automated, ITOps teams obtain the bandwidth to do the kinds of tasks machines are traditionally unable to do, including creatively solving problems, upgrading existing technologies, and planning for the future.

The blistering rate of change in technology and customer needs is driving your business to transform from a heavily device-centric to business service–centric management approach for driving intelligent automation. Traditional IT approaches are keeping IT organizations from understanding the true impact of performance issues on customers and the business. It's difficult to remain a leader in highly competitive markets without a clear view of the risk of changes required to deliver solutions fast enough to stay ahead of your business and customers' expectations.

ScienceLogic offers four solutions that seamlessly adapt to your evolving IT operations' needs. Whether you are looking to consolidate your infrastructure monitoring, automate your incidents, optimize service health for your critical applications and infrastructure, or you need to leverage machine learning to scale for growth—we've got you covered.

Business Service Management

Business service management (BSM) is a strategic approach that has transformed the way organizations manage their IT services. It represents a paradigm shift from the traditional IT-centric view and toward a holistic perspective that aligns IT services with the broader goals and objectives of the business.

But on the surface, this might seem a bit obvious. Align IT services with the broader objectives of the business? When was this ever *not* the goal?

As IT evolved over the past several decades, executives have often found themselves out of sorts with the proliferation of new devices and tools that seemed to be adopted solely for the sake of technology itself rather than as a means to enhance business outcomes. In the early days of IT, technology adoption was driven more by a desire to keep up with the latest trends rather than a clear understanding of how it could directly benefit the business.

The advent of personal computers, the internet, mobile devices, and a multitude of software applications created a complex and rapidly evolving IT landscape. In this environment, IT departments often became siloed, each responsible for a specific aspect of technology. The result was a lack of cohesion between IT and the rest of the organization, making it challenging to ensure that technology investments were genuinely contributing to business success.

This lack of alignment became more apparent as organizations encountered issues such as:

1. **Overwhelming Complexity:** The proliferation of IT tools and services led to complexity that was difficult to manage. Organizations found themselves with disparate systems that didn't communicate effectively, leading to inefficiencies and increased operational costs.

2. **Downtime and Disruptions:** The lack of holistic oversight meant that IT incidents and outages were common. These disruptions could grind business operations to a halt, resulting in financial losses and damage to the organization's reputation.

3. **Misallocation of Resources:** Without a clear understanding of how technology impacted business processes, organizations often misallocated resources.

They invested in technology that didn't align with their core objectives while neglecting critical areas.

4. **Inadequate Customer Experience:** The focus on technology for technology's sake often led to a disconnect between IT services and the end-user experience. Customers and employees faced usability issues and frustrations, impacting satisfaction and productivity.

5. **Inefficiencies and Cost Overruns:** IT projects frequently suffered from scope creep and cost overruns due to a lack of visibility into how they impacted the broader business. This eroded the return on investment for technology initiatives.

BSM emerged as a response to these challenges. It introduced a fundamental shift in mindset for both corporate decision-makers and IT managers alike, emphasizing the need to view IT as a provider of services that directly influence business processes and outcomes. It sought to bridge the gap between technology and business by:

1. **Prioritizing Business Objectives:** BSM ensures that technology initiatives are chosen and executed based on their potential to enhance specific business objectives. It moves beyond technology for technology's sake and focuses on meaningful outcomes.

2. **Ensuring End-to-End Visibility:** BSM provides a holistic view of IT services, enabling organizations to understand the entire service delivery chain. This

visibility facilitates proactive issue resolution and efficient resource allocation.

3. **Measuring Business Impact:** BSM emphasizes the measurement of key performance indicators (KPIs) and service-level agreements (SLAs) to gauge how IT services contribute to business success. It links IT performance directly to the achievement of strategic goals.

5. **Enhancing Collaboration:** BSM encourages collaboration between IT teams and other business units. It promotes a shared understanding of how IT supports business functions, fostering better decision-making and teamwork.

In essence, BSM isn't just about aligning IT with business goals. It's also about ensuring that IT investments and services are meaningful contributors to an organization's success. It's a strategic approach that recognizes the vital role of IT in today's business landscape and leverages it effectively to gain a competitive edge, drive efficiency, and improve the overall customer experience.

The evolution of BSM can be traced through several key phases:

Pre-2000s: IT Component Management

In the era preceding the 2000s, the landscape of IT management was characterized by a paradigm that revolved around the management of individual components within an organization's technology infrastructure. The central tenet of IT management during this period was a *component-centric mindset*. IT professionals primarily concerned themselves with the discrete elements that made up the technology stack, such as servers, routers, workstations, and software applications. Each of these components was managed as an isolated entity, with its own set of responsibilities and maintenance routines.

But one of the critical challenges of this component-centric approach was the prevalence of siloed IT teams. Different teams within an organization often managed specific components, leading to a fragmented and compartmentalized IT environment. For example, networking teams managed the routers and switches, while application teams were responsible for software development and maintenance. These silos could hinder effective communication and collaboration across teams.

And perhaps the biggest drawback of this approach was the limited visibility into the broader impact of individual components on the organization's business processes. IT managers had a narrow view of their technology landscape, which made it challenging to gauge how changes or issues in one component could ripple through the entire system. This lack of holistic insight could lead to unexpected disruptions and downtime. With a focus on individual components, IT management often adopted a *reactive stance*. Teams would typically respond to incidents and problems as they arose, troubleshooting and fixing issues on a case-by-case basis. This approach could result in a perpetual cycle of firefighting rather than proactive problem prevention.

During this era, maintaining and upgrading individual components demanded substantial resources. Organizations invested heavily in hardware and software, and the lifecycle management of these components required ongoing attention.

This placed a significant burden on IT budgets and personnel. And so as businesses grew and evolved, the component-centric approach struggled to scale efficiently. Adding new components or integrating them into existing systems often proved to be complex and time-consuming, which could hinder an organization's ability to adapt to changing market conditions.

Security was a concern in this era as well, but the security measures were often fragmented and lacked the comprehensive strategies we see today. With limited visibility into the interconnectedness of components, it was challenging to develop—let alone enforce—robust security policies.

It's important to recognize that this component-centric approach was not without its successes. It laid the groundwork for many of the technologies and practices we rely on today. However, as technology and business environments became increasingly complex and interconnected, it became evident that a more holistic and integrated approach to IT management was needed.

This realization paved the way for the transformation of IT management practices in the post-2000 era.

Late 1990s–Early 2000s: Emergence of BSM

The turn-of-the-century era marked a pivotal period in the evolution of IT management practices, with the emergence of business service management (BSM). Several interconnected factors drove the adoption of BSM, transforming the way organizations approached technology management and its alignment with business objectives.

During this era, IT environments were becoming increasingly complex. Organizations were expanding their technology infrastructure with a multitude of interconnected components, including servers, networks, databases, and applications. This complexity presented challenges in terms of monitoring, managing, and ensuring the reliability of these systems.

The late 1990s also witnessed a massive surge in e-commerce and online business activities. Companies began to rely heavily on digital platforms to engage with customers, to conduct transactions, and to deliver services. The success and competitiveness of these e-commerce ventures were intrinsically tied to the performance and availability of IT systems.

And as businesses became more digitally driven, they grew increasingly dependent on IT for their day-to-day operations. Downtime or disruptions in IT services directly translated to financial losses, damaged reputation, and dissatisfied customers.

Recognizing this, organizations needed a more robust approach to ensure the reliability and performance of their IT systems.

BSM emerged as a response to these challenges. It aimed to bridge the gap between IT and the business by shifting the focus from managing individual IT components to managing the services and processes that IT enabled. BSM sought to align IT services with the strategic objectives and needs of the organization. BSM introduced a more holistic perspective on IT management. It encouraged organizations to view IT as a collection of interconnected services that directly impacted business outcomes. This approach promoted a better understanding of how IT services contributed to revenue generation, customer satisfaction, and overall business success.

BSM was also closely linked to the adoption of service-oriented architecture (SOA), which promoted the development and delivery of IT services as modular and reusable components. SOA facilitated the creation of flexible, agile, and business-responsive IT systems. The adoption of BSM was facilitated by the development of tools and technologies that allowed organizations to monitor and manage IT services comprehensively. These tools provided real-time insights into service performance and helped identify and address issues proactively.

BSM aimed to shift IT management from a component-centric approach to a service-centric one, with a focus on aligning IT services with business objectives. This shift laid the foundation for modern IT service management practices, emphasizing the importance of understanding and delivering IT services that drive business value.

2000-2009: Maturation of BSM Solutions

During the first decade of this century, BSM continued to evolve and reached what's arguably a critical stage of maturity. This period saw the emergence of dedicated BSM solutions and platforms that played a pivotal role in transforming how organizations managed their IT environments.

By around 2005, organizations had come to recognize the significant value that BSM brought to the table. It wasn't just a theoretical concept anymore; it had become a practical necessity. BSM offered a strategic approach to IT management, aligning technology with business goals and ensuring that IT services delivered real value. And so BSM solutions were increasingly designed to work in tandem with existing IT management tools and systems. These solutions integrated with various components of an organization's IT infrastructure, such as network monitoring, application performance management, and service desk software. This integration facilitated a seamless transition to a BSM-oriented approach without disrupting existing operations.

One of the hallmark features of BSM solutions was their ability to provide a unified view of IT services. These platforms offered a comprehensive perspective on how IT services impacted the broader business landscape. They aggregated data from diverse sources to present a consolidated view of service health, performance, and their alignment with business processes. BSM solutions introduced real-time monitoring and analysis capabilities. They allowed organizations to track the performance of critical IT services and processes in real time. This proactive monitoring enabled faster issue detection, reducing downtime and minimizing the impact on business operations.

BSM solutions went beyond technical metrics and introduced business-relevant key performance indicators (KPIs). These metrics provided insights into how IT services influenced key business outcomes, such as revenue generation, customer satisfaction, and operational efficiency. This shift in focus reinforced the notion that IT wasn't just a cost center, but it was also a strategic asset. BSM platforms incorporated advanced root cause analysis tools. When issues arose, these tools helped IT teams pinpoint the exact source of problems, reducing the time and effort required for troubleshooting. This proactive approach was crucial for maintaining service reliability.

BSM solutions also embraced automation and orchestration capabilities. They allowed organizations to automate routine IT tasks and orchestrate complex workflows, enhancing efficiency and reducing manual intervention. So as organizations continued to grow, BSM solutions offered scalability and flexibility to accommodate changing IT landscapes. They could adapt to new technologies, services, and business processes, ensuring that the BSM framework remained relevant.

2010s–Today: Modern BSM

From roughly 2010 to the present day, BSM has seen continuous evolution into a modern and indispensable practice for organizations striving to optimize their IT operations and align them with evolving business objectives. During this era, BSM solutions have incorporated automation and other critical features that have revolutionized the way IT services are managed.

One of the standout features of modern BSM is the integration of *advanced automation and orchestration capabilities.* Automation has become a linchpin of IT operations, enabling the swift execution of such routine tasks as software updates, server provisioning, and incident remediation. Orchestrating complex workflows across the IT landscape ensures that processes are executed seamlessly, minimizing manual intervention and reducing the risk of errors.

Modern BSM solutions excel at speeding up incident resolution. They use automation to detect issues in real time, allowing for immediate response and remediation. This proactive approach minimizes service disruptions and ensures higher availability, which is crucial in today's 24/7 business environment.

To ensure that IT services meet business expectations, modern BSM incorporates the use of SLOs and SLIs. These metrics define the desired service performance levels and the indicators that track them. By continuously monitoring these indicators and comparing them to predefined thresholds, organizations can quickly identify deviations and take corrective actions to maintain service reliability.

Modern BSM leverages predictive *analytics and artificial intelligence* to foresee potential issues and optimize IT service performance. Machine learning algorithms analyze historical data to identify patterns and anomalies, enabling proactive measures to prevent incidents before they occur. AI-powered chatbots and virtual agents also enhance customer support and incident management.

And in the era of cloud computing and hybrid IT environments, modern BSM has adapted to manage services that span on-premises and cloud-based infrastructure seamlessly. It ensures consistent service delivery and performance across these diverse environments while allowing organizations to take full advantage of the scalability and flexibility offered by the cloud.

Modern BSM embraces DevOps principles, fostering collaboration between development and operations teams. This alignment ensures that changes to IT services are seamlessly integrated, tested, and deployed while maintaining service stability and performance.

The practice of continuous improvement is integral to modern BSM. It involves regularly reviewing and optimizing IT processes, services, and performance based on data-driven insights. This iterative approach helps organizations stay agile and responsive to changing business demands.

Security has become paramount, and modern BSM incorporates robust security measures. BSM solutions monitor and enforce security policies, quickly identifying and mitigating threats to maintain IT services' confidentiality, integrity, and availability.

Lastly, modern BSM places a strong emphasis on customer satisfaction. It recognizes that IT services exist to support the needs of the business's and its customers' needs. Organizations can enhance their competitiveness and reputation by aligning IT services with customer expectations and business outcomes.

In summary, the evolution of BSM from the 2010s to the present day has been marked by the integration of automation, predictive analytics, AI, and other advanced capabilities. BSM has become an essential practice for organizations looking to maximize the value of their IT investments, enhance customer satisfaction, and adapt to the ever-changing IT and business landscapes. It is a dynamic and evolving field that continues to shape the way businesses leverage technology to achieve their objectives.

So what's next for BSM? How can those in leadership positions guarantee that the ongoing and rapidly intensifying expansion of technology is seamlessly integrated with the overarching objectives of their organizations?

The answer lies in AI—something we noted above, but it has yet to realize its full potential in IT operations. Here are a few ways to think about how AI can (and will) shape the future of BSM.

1. **Predictive Analytics and Prescriptive Actions:** AI can take predictive analytics to the next level. By analyzing vast datasets and identifying subtle patterns, AI algorithms can predict IT service issues with greater accuracy and foresee potential bottlenecks or disruptions before they impact business operations. Moreover, AI can provide prescriptive actions to IT teams, recommending optimal strategies to prevent or mitigate problems, allowing for proactive management.

2. **Autonomous IT Operations:** The future of BSM may involve autonomous IT operations driven by AI. AI-powered systems can autonomously manage routine tasks, perform real-time monitoring, and respond to incidents without human intervention. This not only enhances operational efficiency but also reduces the risk of human error.

3. **Advanced Root Cause Analysis:** AI can significantly improve root cause analysis. Instead of relying on rule-based systems, AI can analyze complex relationships across IT components, identifying the true source of issues. This capability accelerates problem resolution and minimizes the impact on business processes.

4. **Cognitive Service Desks:** AI-driven cognitive service desks can transform customer support and incident management. Chatbots and virtual agents can provide immediate assistance to users, answer common queries, and guide them through issue resolution. This not only improves customer satisfaction but also frees up IT staff to focus on more complex tasks.

5. **AI-Enhanced Security:** Cybersecurity is a top concern, and AI can play a pivotal role in enhancing an organization's security posture. AI systems can continuously analyze network traffic, detect anomalies, and respond to threats in real-time. They can also automate security patching and updates to stay ahead of vulnerabilities.

6. **AI-Powered Business Insights:** AI can provide deeper insights into how IT services impact business outcomes. By correlating IT performance data with business metrics, organizations can gain a better understanding of the financial, operational, and customer implications of their IT investments. This information is invaluable for strategic decision-making.

7. **AI-Driven Cost Optimization:** AI can help organizations optimize their IT spending. AI algorithms can analyze cost data and recommend cost-saving measures, such as resource consolidation, right-sizing infrastructure, and identifying underutilized assets.

8. **Continuous Learning and Adaptation:** AI systems can continuously learn from data and adapt to changing IT and business environments. This adaptability ensures that BSM practices remain relevant and effective in the face of evolving technologies and market dynamics.

9. **Ethical Considerations:** As AI becomes more integral to BSM, organizations must also consider ethical and responsible AI practices. Ensuring that AI algorithms are fair, transparent, and bias-free is critical to maintaining trust and compliance.

It's indisputable: AI holds immense promise for the future of BSM. It has the potential to revolutionize IT operations entirely. But to unlock this potential, organizations must invest in AI technologies, develop the necessary expertise, and adhere to ethical guidelines. As technology continues to evolve, AI-powered BSM will play a pivotal role in helping organizations navigate the complexities of the digital age and remain competitive in their respective industries.

11

Why ScienceLogic?

BEING AT THE center of a revolution is not an uncommon place for ScienceLogic. The team at ScienceLogic have always been frontier builders and continue to be even now during the AIOps revolution. As a pioneer in the IT operations space, we support IT operations for thousands of the world's biggest brands, implementing and creating solutions on the cutting-edge of technological innovation. And nothing is more exciting than being on the cutting-edge of the burgeoning AIOps revolution. How can ScienceLogic support your organization as it moves into the next transformative phase of technology? Selecting ScienceLogic as your solution for managing on-premises and hybrid cloud application environments offers several compelling advantages. Below are pivotal reasons why ScienceLogic's AIOps platform stands out.

Comprehensive Visibility

With ScienceLogic, you'll achieve a level of visibility that transcends the boundaries of traditional IT management. This platform provides a holistic view of your complete IT infrastructure, whether it resides on-premises, in the cloud, or in a hybrid environment. From a single unified platform, you'll gain invaluable insights into the intricate web of your IT landscape. This includes granular insights, real-time monitoring, historical data, cross-platform integration, and customizable dashboards, all of which empower you to make informed decisions, proactively address issues, and ensure the optimal performance of your entire IT ecosystem.

Superior AIOps Capabilities

Unlocking the Potential of Artificial Intelligence for IT Operations (AIOps), ScienceLogic introduces a new era of efficiency in your IT management. With its advanced AIOps capabilities, this platform transforms the way you handle monitoring, analysis, and response within your IT infrastructure. It leverages AI to provide automated insights, proactive issue resolution, data-driven intelligence, root cause analysis, predictive analytics, and enhanced resource allocation. This not only saves time but also ensures that critical information doesn't get lost in the noise, and empowers you to focus on strategic initiatives rather than firefighting daily IT challenges.

A Unified Platform

Say goodbye to the daunting challenge of juggling numerous tools to handle a myriad of tasks. With ScienceLogic, you can seamlessly integrate monitoring, analytics, and automation into a single, cohesive platform, streamlining your overall management process and eradicating the headaches that come with tool sprawl. No longer will you need to grapple with the confusion and inefficiency that arises from using multiple, disjointed tools to keep track of your systems and processes. ScienceLogic offers a holistic solution that consolidates all your monitoring needs, provides insightful analytics, and empowers you with automation capabilities—all within one user-friendly interface. This unification of tools not only simplifies your workflow but also enhances your productivity. Imagine having a centralized command center where you can

Hybrid Cloud Support

The hybrid cloud paradigm, characterized by its blend of on-premises infrastructure and cloud resources, presents unique challenges when it comes to managing applications and ensuring their optimal performance. ScienceLogic deftly rises to meet these challenges, offering a comprehensive solution that seamlessly monitors and manages applications across a diverse array of environments, encompassing both on-premises and cloud-based systems. This exceptional capability guarantees that organizations can maintain a steadfast commitment to delivering a consistent and dependable user experience. By leveraging ScienceLogic, businesses can harmoniously orchestrate the monitoring and management of applications, regardless of where they reside. Whether it's a mission-critical application housed within the organization's data center or a dynamic, cloud-native application hosted on a public cloud platform, ScienceLogic's capabilities extend across this complex landscape, offering a unified lens through which to oversee and govern these applications.

Security and Compliance

In a world where data breaches and cyber threats are constant concerns, ScienceLogic's security features provide multifaceted defense mechanisms. They enable you to proactively identify vulnerabilities, assess risks, and implement proactive measures that align with the specific requirements of your industry and regulatory mandates. ScienceLogic's extensive suite of security features serves as a formidable ally in your organization's ongoing commitment to maintaining compliance with the ever-evolving landscape of industry standards and regulations. These powerful tools not only bolster your capacity to adhere to established guidelines but also enhance your ability to effectively monitor security incidents and safeguard the integrity of your data. By leveraging ScienceLogic's security capabilities, you gain a comprehensive view of your organization's security posture. This encompasses real-time monitoring of critical systems, networks, and applications, ensuring that any irregularities or potential threats are promptly detected and addressed. Moreover, ScienceLogic's advanced analytics and reporting tools allow you to analyze security data with precision, enabling you to make informed decisions and swiftly respond to emerging threats.

Cost Optimization

ScienceLogic takes your IT management to the next level by going above and beyond to optimize the usage of your valuable resources and taking proactive measures to address issues before they escalate. This commitment leads to a twofold benefit:

significantly reduced operational costs and the maximization of your return on investment for your IT assets.

Efficient resource usage is at the core of ScienceLogic's mission. The platform continuously assesses the performance and availability of your IT assets, whether they are servers, storage devices, or networking components. Through real-time monitoring and advanced analytics, ScienceLogic identifies opportunities to streamline resource allocation, ensuring that your infrastructure is used optimally. This means you're not wasting money on underutilized hardware or over provisioned resources, allowing you to get the most out of your IT investments. Also, instead of waiting for problems to arise and disrupt operations, ScienceLogic anticipates potential issues through predictive analysis and early warning systems. This foresight enables your IT teams to take corrective action before users are impacted, preventing costly downtime and service disruptions. By nipping problems in the bud, ScienceLogic helps you avoid the high operational expenses associated with reactive IT support.

Proactive Problem Resolution

ScienceLogic takes proactive IT management to a whole new level. By harnessing the power of AI and machine learning, ScienceLogic not only identifies potential issues but also predicts and resolves them before they can disrupt your users, thus guaranteeing a seamless and uninterrupted user experience. IT downtime or performance hiccups can be costly and frustrating. ScienceLogic acts as a vigilant guardian, constantly monitoring your IT environment, whether it's on-premises or in the cloud. Through its AIOps capabilities, the platform ingests vast amounts of data from various sources, including performance metrics, logs, and historical data. This wealth of data is then analyzed and processed by advanced AI algorithms that have been trained to recognize patterns, anomalies, and potential trouble spots. ScienceLogic's AI not only spots existing issues but also has the capacity to forecast emerging problems based on historical trends and patterns. This predictive capability allows you to take preemptive action to avert disruptions. The result is a user experience that remains consistently smooth and uninterrupted. Your users won't have to contend with the frustration of slow applications, inaccessible services, or unexpected outages. By leveraging ScienceLogic's AIOps capabilities, you can provide your users with the reliability and quality of service they expect in today's digital age.

Support and Expertise

When you choose ScienceLogic, you're not just investing in a powerful technology solution; you're gaining access to a world-class customer support experience that is second to none. Count on ScienceLogic to provide you with comprehensive customer support and access to a team of seasoned experts who are ready and eager to assist you with every aspect of your journey, from initial setup and configuration to troubleshooting and ongoing optimization.

1. **Initial Setup:** Getting started with a new technology solution can often be a daunting task. ScienceLogic

recognizes this and is committed to making your onboarding experience as smooth as possible. The expert support team will guide you through the setup process, helping you configure the platform to align perfectly with your unique business needs. They'll ensure that you're up and running efficiently from day one.

2. **Troubleshooting:** In the dynamic world of IT management, challenges and issues can arise unexpectedly. When they do, ScienceLogic's support team is your first line of defense. These experts are well-versed in diagnosing and resolving a wide range of technical issues. Whether it's a sudden performance dip, an anomaly in your data, or any other IT-related problem, you can rely on ScienceLogic's support to swiftly identify the root cause and provide effective solutions.

3. **Optimization:** Technology is ever-evolving, and to stay competitive, your IT environment must continually adapt and improve. ScienceLogic's support extends beyond solving immediate problems; it includes ongoing optimization guidance. The experts will work closely with you to ensure that you are fully leveraging the platform's capabilities to their maximum potential. This proactive approach helps you extract more value from your investment, improve efficiency, and enhance your overall IT operations.

4. **Continuous Learning:** ScienceLogic's support is not just about fixing issues; it's also about empowering you with knowledge. The support team provides valuable insights, best practices, and training resources to help you and

your team become proficient users of the platform. This commitment to continuous learning ensures that you're always equipped with the skills and knowledge needed to make the most of ScienceLogic.

5. **Timely Responses:** When you reach out to ScienceLogic's support, you can expect prompt and attentive service. The team understands that downtime and disruptions can be costly, and they are dedicated to resolving your concerns in a timely manner, minimizing any impact on your operations.

ScienceLogic's customer support is not just a service; it's a partnership. It's a commitment to being by your side throughout your IT management journey, offering guidance, solutions, and expertise to ensure your success. With ScienceLogic, you're not just purchasing a product; you're investing in a holistic experience that empowers your organization to thrive in the world of IT management.

Ecosystem Integration

ScienceLogic takes the concept of integration to a new level by offering a seamless and adaptable solution that effortlessly harmonizes with a wide range of tools and platforms, providing robust support for your existing infrastructure and toolchain. This inherent flexibility and future-ready design make ScienceLogic the perfect choice to accommodate your organization's expansion plans and evolving needs. We're designed to break down silos and promote interoperability, seamlessly integrating with a multitude of existing tools and platforms that your organization relies on, whether they are monitoring solutions, ticketing systems, or cloud management platforms. This integration ensures that your existing investments are leveraged to their fullest potential, creating a unified ecosystem where data flows seamlessly between systems. We also offer cross-platform compatibility, meaning that regardless of your existing infrastructure, you can trust ScienceLogic to provide comprehensive monitoring and management capabilities. Finally, as your organization grows and your IT environment expands, ScienceLogic grows with you. Its scalability ensures that it can handle the increased volume of data, devices, and applications without missing a beat. Whether you're adding more servers, adopting new cloud services, or expanding your network, ScienceLogic can scale to accommodate your evolving requirements.

In summary, ScienceLogic stands as a dependable choice for managing on-premises and hybrid cloud application environments due to its cutting-edge AIOps capabilities, scalability, comprehensive visibility, and capacity to streamline IT operations while simultaneously reducing risks and costs. It empowers organizations to elevate their IT performance, ensuring a superior experience for both users and customers alike.

12

The Core Reason Organizations Choose ScienceLogic

ENTERPRISES ARE RAPIDLY turning to more hybrid cloud architectures and leveraging a broader mix of software to power business services. As a result, IT operations teams must contend with an increasingly complex web of interconnected services that must be monitored, maintained and optimized to meet business SLAs for superior customer experiences. On top of this, there is a scarcity of available IT talent skilled in supporting rapidly changing technology stacks used to power new digital services.

To support new digital business imperatives faster while delivering a reliable customer experience, businesses need a new approach. ScienceLogic's intelligent AIOps platform creates an autonomous IT estate that can maintain itself while providing staff with recommendations on how to improve IT performance. ScienceLogic accelerates innovation, reduces IT costs and eliminates risk through:

- Future-proof data acquisition and intelligent noise reduction
- Human-friendly root cause and predictive analysis via AI reasoning and correlated telemetry

- AL-Drien recommendations to enhance IT using organizational knowledge
- Explainable and auditable autonomous execution

Take Your AIOps Strategy to the Next Level

Our AIOps platform enables new levels of operational efficiency and business service availability through automated, ML-driven issue remediation and recommendations on how to optimize operations. And with extensible monitoring that can support new digital technologies faster, you empower rapid innovation without compromising SLAs. With ScienceLogic your IT estate becomes a completely connected, 'living' organism. It continuously scans your entire ecosystem in detail to predict and resolve issues automatically while recommending changes to improve the performance of your business. There are several advantages to this, including:

1. **Maximize your return on investments and reduce cost overruns:** Continual resource monitoring enables IT budget owners to ensure resources are being used as effectively as possible and better plan for new purchases and manage cloud costs. As IT devices and services reach maximum usage, you can automatically allocate resources incrementally so the business continues to run smoothly while also optimizing deployment size.

2. **Deliver superior customer experiences and minimize downtime:** Our end-to-end incident resolution combines ML-driven root cause analysis with automated

repair to minimize downtime. Predictive analysis identifies anomalies and provides recommendations, in an easy to understand format, so issues can be resolved well before they disrupt business operations. Combined, your infrastructure is able to repair and maintain itself so there is less risk of disruption to customer experiences.

3. **Optimize and leverage your existing IT talent:** Through the use of large language models our AI provides insights and recommendations that are human-friendly and easily understood by most of your organization. This enables even generalists to take action and ensures the precious skills or your most qualified staff are available for more fulfilling projects like innovation rather than tied up with mundane tasks.

4. **Align IT and business leaders for success:** We help organizations agree to and achieve SLAs based on a common view of how the business runs. By correlating hybrid cloud resources into a single business service view, we simplify management and maintenance. When issues occur, our platform automatically identifies root causes and notifies IT as well as business teams to ensure responses are coordinated and aligned.

5. **Accelerate innovation and increase IT agility:** Our platform can easily be extended to support new systems and software, even less common resources like 5G towers, so that your business can continually adopt new technologies without creating blind spots in your operations strategy.

Intelligent, AI for IT Operations
That Few Can Match

With comprehensive IT asset monitoring and integrated AI and automation, ScienceLogic delivers a single platform that goes beyond AIOps as it exists today. We bring you intelligent, easy to understand insights that empower you to enhance – not just maintain – even the most complex ecosystems. We identify and predict emerging issues before they can impact the business, eliminate human error, and enable you to make the most of your valuable IT investments. At the core, our approach to AIOps delivers an unrivaled ability to observe, engage, and act to make you:

1. **More agile and complete monitoring of your business:** Our adaptive and extensible approach to data acquisition provides more comprehensive asset discovery and extraction of data so you can innovate faster while reducing blind spots that can cause outages. In fact, we are proven to enable 100% CMDB accuracy for better IT estate management. We also provide intelligent noise reduction to remove duplicates and false positives, enabling faster analysis.

2. **More insightful and easily understood:** Our generative AI leverages telemetry data and contextual information to uncover issues, predict future events and provide a more accurate diagnosis of root cause. This is combined with LLM driven recommendations that are easily understood by more of your staff. Combined, this takes

you beyond technical 'break-fix insights' to easily action-able recommendations.

3. **More autonomous:** Integrated AI and automation enables you to go beyond basic troubleshooting and ticket creation to reliable, automated issue management. We let you automate more of your operational tasks and give you greater control so you can decide what is fully automated, and what is AI-guided. This gives you a greater range of choice in terms of automation of ITSM, CMDB, DevOps, compliance, and other requirements. These deep differentiators mean our AIOps platform allows you to realize the vision of an autonomous IT stack that supports ongoing digital business innovation, without forcing you down a specific path that might have more limited control and functionality. And you can rely on it because it works with you. We also stand out because we are a mature, highly extensible platform built for real IT— our platform brings a deep understanding of IT and an expert perspective on how to harness AI/ML, automation, and more to further business goals. We offer the most capable AIOps support in a single plat-form—no other single platform is as inherently scalable, sustainable, robust, and extensible for modern enter-prises as ours.

Built for today, ready for tomorrow, ScienceLogic is your ideal choice for enhancing and automating observability, problem detection, IT optimization and response across complex busi-ness ecosystems. By integrating ScienceLogic into their cloud infrastructure, we have helped clients improve productivity by 85% and reduce incidents by 66%. Our advanced AIOps platform continuously brings your IT operations teams broader visibility to how the business is operating and automates a broader range of IT tasks to minimize risk and deliver superior business outcomes. By moving you toward a business-first architecture, we enhance productivity, reduce costs, accelerate innovation and close security gaps while focusing your IT talent on high-value tasks. And because our platform is built to handle tomorrow's hybrid cloud infrastructures today, we can help drive your business steadily toward the future with confidence.

www.ingramcontent.com/pod-product-compliance
Lightning Source LLC
Chambersburg PA
CBHW021929190326
41519CB00009B/963